MAMMALS
of the
NATIONAL PARKS

JOHN H. BURDE *and* GEORGE A. FELDHAMER

THE JOHNS HOPKINS UNIVERSITY PRESS
Baltimore and London

Printed in China on acid-free paper
9 8 7 6 5 4 3 2 1

The Johns Hopkins University Press
2715 North Charles Street
Baltimore, Maryland 21218-4363
www.press.jhu.edu

Library of Congress Cataloging-in-Publication Data

Burde, John H., 1946 –
  Mammals of the national parks / by John H. Burde and George A. Feldhamer.
     p. cm.
  Includes index.
  ISBN 0-8018-8097-1 (hardcover : alk. paper)
  1. Mammals—United States. 2. National parks and reserves—United States.
I. Feldhamer, George A. II. Title.
  QL717.B857 2005
  599'.0973—dc22   2004014272

A catalog record for this book is available from the British Library.

Book design by Bill Marr

Photography and illustration credits may be found on page 212,
which is an extension of this copyright page.

Frontispiece: A coyote howls in the snow.
Title page photograph: A grizzly bear passes moose antlers
on the shore of Naknek Lake in Katmai National Park.

To Beverly, Joel, and Jared. —JHB

To Carla, Andy, and Carrie. —GAF

AMERICA'S NATIONAL PARKS

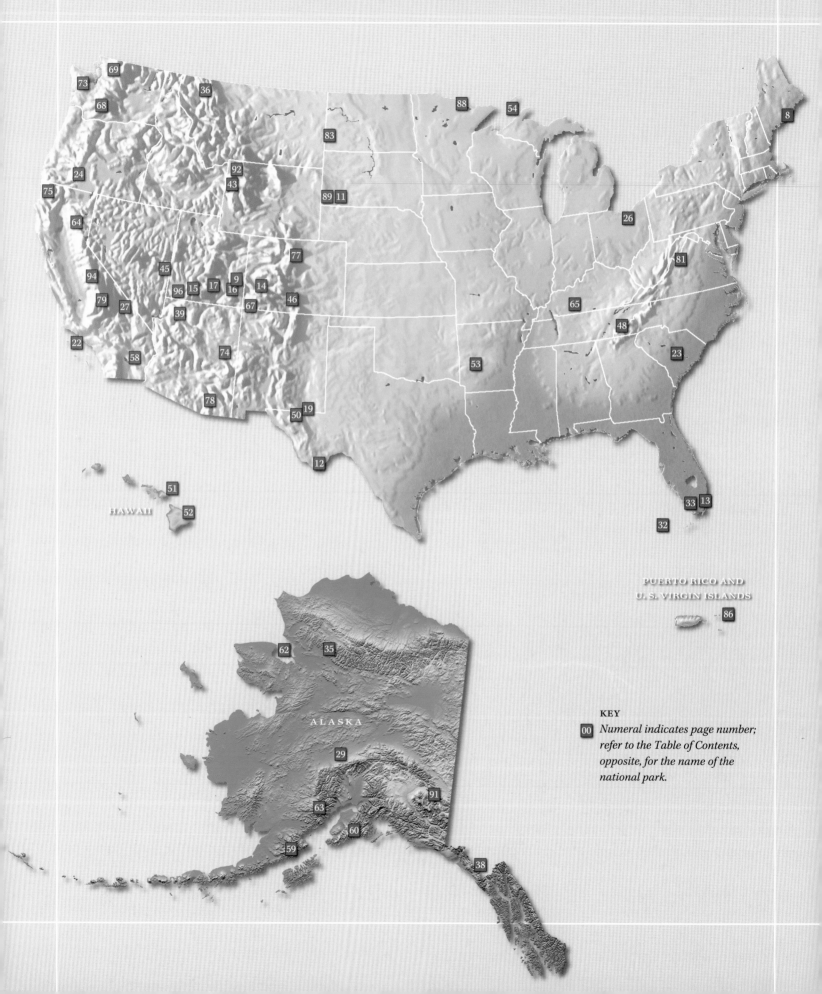

HAWAII

PUERTO RICO AND
U. S. VIRGIN ISLANDS

ALASKA

KEY

00 *Numeral indicates page number;
refer to the Table of Contents,
opposite, for the name of the
national park.*

CONTENTS

One of America's respected writers, Wallace Stegner, liked to remark that national parks were the best idea America ever had. Indeed, the national park idea was an American contribution to the world. Today, more than one hundred countries have created national parks. Additional countries continue to join the list of those who have preserved their natural and cultural heritage for future generations to enjoy. Scotland became the most recent country to establish a national park when Loch Lomond and The Trossachs National Park was created in 2002.

The idea of national parks in the United States can be traced to George Catlin, perhaps best noted for his preservation of Native American culture on the Great Plains through his paintings. Catlin became concerned whether indigenous cultures could survive in the face of extraordinary growth and expansion. He proposed that large expanses of prairie be preserved for Native American cultures as "a nation's park, containing man and beast, in all the wildness and freshness of their nature's beauty."

In 1870, an expedition led by Henry Washburne, Territorial Surveyor-General of Montana, Nathaniel Langford, businessman and politician, and lawyer Cornelius Hedges entered the Yellowstone region. Hedges later wrote: "I think a more confirmed set of skeptics never went out into the wilderness than those who composed our party, and never was a party more completely surprised and captivated with the wonders of nature." At their campsite at the confluence of the Gibbon and Firehole Rivers, under what is now called National Park Mountain, the party concluded that Yellowstone should be preserved.

A porcupine feeds on pussy willows.

News of the expedition created considerable attention. The following year an official expedition, led by Ferdinand V. Hayden, scientifically surveyed Yellowstone. Accompanying the expedition were pioneer photographer William Henry Jackson and artist Thomas Moran. Hayden's words, Jackson's photos, and Moran's art became the basis for a proposal to set Yellowstone aside as a national park. In 1872, the Congress of the United States established Yellowstone as the world's first national park.

The idea of national parks became reality and soon began to grow. The world's second national park, Royal National Park, was established south of Sydney, Australia. Banff, Canada, soon followed. Back in the United States, Sequoia National Park was established in 1890, followed a few days later by Yosemite and General Grant National Parks. The most recent addition to the list is Colorado's Great Sand Dunes National Park, which became America's newest national park.

National parks are special places. They protect a nation's natural and cultural heritage. Not only do parks protect our most beautiful places, they also preserve the ecosystems in those places and the plants, animals, and fossils they contain.

Research has shown that the presence of wildlife is one of the most important reasons people visit national parks. Parks contain a multitude of habitats supporting wildlife of all kinds: birds, mammals, reptiles, amphibians, fish, and innumerable invertebrates. Many threatened and endangered species occur in parks, which provide needed refuge for their survival.

Mammals are the big attraction in most parks. Seeing them heightens a visitor's experience. Some mammals, like chipmunks and squirrels, are observed by most visitors, but seeing a mountain lion is an event only extremely fortunate visitors will experience. Observing a particular mammal is a function of chance, size, and behavior of the species. Knowing when and where to look certainly enhances the prospect of success. This book not only reveals diversity of mammalian species within the national parks but also helps park visitors better understand some of the best places to look for these amazing animals.

The book is divided into two sections. The first section gives a brief history and description of each of America's national parks. Species or species groups that reside in each park are presented, with information as to where they might best be observed, the probability of seeing them, and information about their abundance and location. Descriptions for each park come from individual park mammal lists and interpretive materials as well as visits to almost every park. Every entry in this section concludes with a discussion of "Conservation Concerns" unique to that park. Congressional designation of a park does not end the need for citizen awareness of the critical conservation issues of each park. Continued vigilance is necessary if we are to protect our parks far into the future.

An elk stands in misty waters in Yellowstone National Park.

The second section of the book describes the mammals. For each species or species group, we provide a brief summary of their geographic distribution, physical characteristics, habitats, feeding habits, and reproductive behavior. A list of parks where each species has been documented or probably occurs is provided. "Conservation Concerns" related to mammalian species are discussed to create awareness about issues affecting populations in the parks and throughout North America. We have made every effort to include all mammals in the national parks based on documented occurrence, lists provided by each park, and known geographic ranges. However, species such as shrews, moles, bats, or rodents are most likely to remain undocumented in a given park because they often are small, cryptic, and nocturnal. If we inadvertently failed to include a mammalian species from a park where it occurs, it was most likely one of these.

We thank the personnel at all the parks who took time to answer our requests for updated information about the current status of mammalian species in each park, current policies, and other information. The invaluable assistance of Lisa Russell of the Environmental Studies Program at Southern Illinois University Carbondale in all phases of this project is gratefully acknowledged. Special thanks also to David Armstrong and John Whitaker for sending annotated lists of mammals and books dealing with several of the parks. We would like to recognize the late mammalogist Richard G. Van Gelder for his 1982 book *Mammals of the National Parks*. Last, we sincerely thank the Oakwood Arts and Sciences Charitable Trust. Without their support this book could not have been possible.

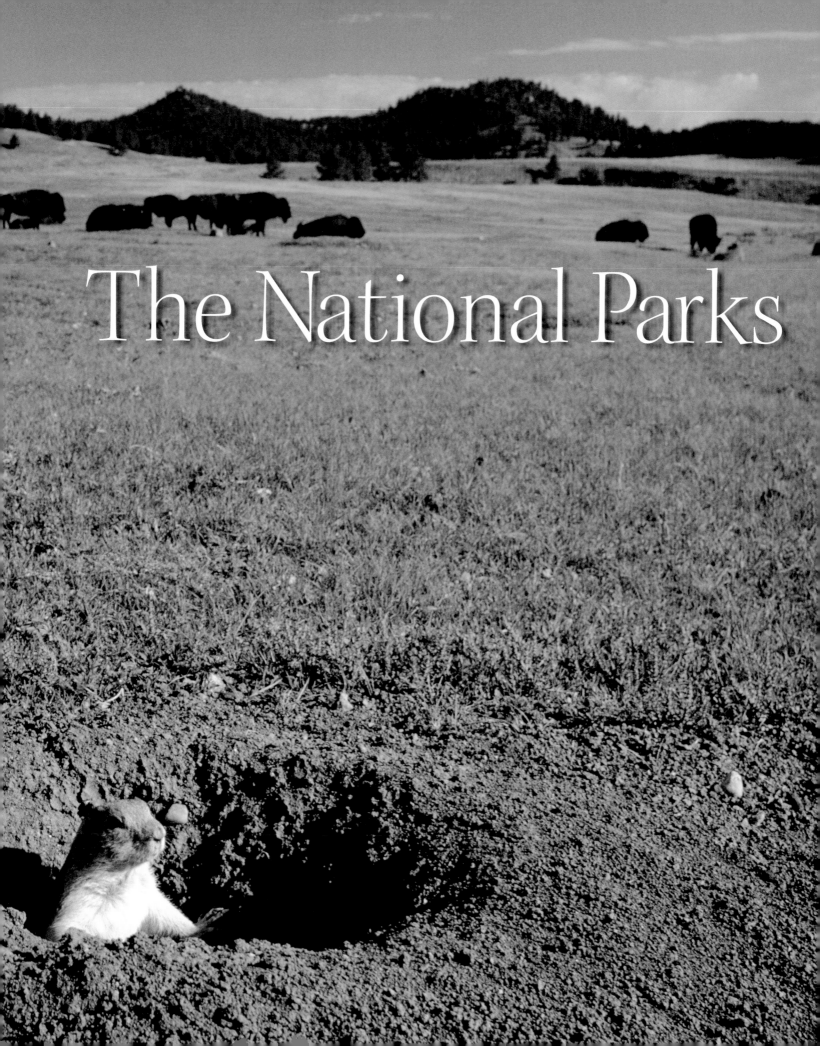

The National Parks

ACADIA

Where the Mountains Meet the Sea

Bar Harbor, Maine

ACRES: 47,400

Once the summer home of America's wealthiest families, Acadia offers views of the Atlantic Ocean as it caresses the rockbound coast. Views from the cliffs occasionally allow glimpses of whales swimming by the park.

Publicly owned ocean frontage is at a premium on the Atlantic coast. There are state parks, and the federal government has established coastal wildlife refuges and national seashores, but there is only one national park on the Atlantic outside south Florida—Acadia National Park in Maine. Located on Mount Desert Island, Acadia protects fifty-two miles of rocky Atlantic shoreline.

In the late nineteenth century Mount Desert Island was the summer home of some of America's wealthiest families, including the Fords, Vanderbilts, Astors, Carnegies, and Rockefellers. Another resident, George Dorr, became so concerned about development on Mount Desert Island that he created a corporation with the purpose of obtaining land to be preserved for future generations. The corporation acquired six thousand acres by 1913; Dorr then offered to donate the land to the federal government for the purpose of creating a national park. In 1916, President Woodrow Wilson accepted the land and declared it Sieur de Monts National Monument. Land acquisition continued on Mount Desert Island; Dorr still hoped for national park status. His hope was realized in 1919 when Lafayette National Park was created, the first national park east of the Mississippi River. It was renamed Acadia National Park in 1929.

The natural environments of Acadia are many and varied. Opportunities abound for visitors to get up close and personal with the sea. Tidal pools teem with life. But there is more to Acadia than the ocean. There are ten freshwater lakes greater than ten acres in the park as well as several smaller ponds. About 10 percent of the park is wetland. Total park acreage is 47,400.

The park is primarily forested with conifers predominating. The most common species include white pine, white and red spruce, balsam fir, eastern hemlock, and northern white cedar. Hardwood species, including aspen, American beech, northern red oak, and several kinds of birch and maple, are also common.

The tops of the mountains are usually treeless granite. In fact, the original name of Mount Desert Island was L'Isles des Monts Déserts, French for the "island of bare mountains," penned by Samuel Champlain in 1604. The tallest mountain is Cadillac Mountain (1,530 feet), the highest point on the Atlantic coast. A few trees, such as pitch pine, brave these windswept heights.

A twenty-seven-mile Park Loop Drive connects several of the park's more noted features along the rugged shoreline and the inland glacial valleys. In Acadia's early days, the automobile was not so universally welcomed. In 1917, John D. Rockefeller, Jr., donated to the park a fifty-seven-mile system of carriage paths that would be closed to autos. These remain closed to motor vehicles today but may be enjoyed by hikers, bikers, horseback riders, and cross-country skiers. Some 120 miles of foot trails provide access to the lakes and forests away from roads.

There are fifty species of mammals at Acadia; most common are the rodents. The mammal perhaps most often observed is the eastern chipmunk. Red squirrels are common throughout the park. Gray squirrels abound, especially in the hardwood forests. They are descendants of squirrels introduced to Acadia in 1922. Both the northern (common) and southern (uncommon) flying squirrels inhabit Acadia, as does the porcupine.

Acadia's riparian areas are home to several mammals including the muskrat. Beaver were extirpated at Acadia early but reintroduced in 1921. They are now common, but their nocturnal behavior reduces the probability of seeing one.

There are several members of the weasel family at Acadia. The short-tailed weasel, long-tailed weasel, mink, river otter, and striped skunk are all common in the park. Even so, they are seldom seen by visitors.

There is a single species of hare in the park, the snowshoe hare. They are nocturnal and may be seen along roads at dusk and dawn.

A major wildlife issue at Acadia is the problem of raccoons in campgrounds, especially Blackwoods. Improper food storage and garbage disposal have made many raccoons campground scavengers. Beggar raccoons must be trapped and destroyed because Maine law prevents relocating animals that may carry rabies.

There are predators in the park. The red fox and the coyote are both common. The bobcat is rare. The black bear is uncommon but is seen from time to time.

The largest mammal you are likely to see is the white-tailed deer. The only other large mammal is the moose. Though not resident in the park, moose may occasionally wander into the park's small lakes and ponds.

A special feature of Acadia is the opportunity to see marine mammals. Many can be seen from tour boats that depart from Bar Harbor and other towns. Whale watcher tour patrons frequently see fin and Minke whales; humpback, right, and pilot whales are less common. Harbor porpoises are occasionally seen as well.

National Park Service concessionaires offer boat tours each summer. Trips to Baker Island provide a close-up look at the common harbor seals that haul up on the rock ledges found along the tour route. Harbor seals can also be seen on Bass Harbor trips. Gray seals are uncommon but are sometimes seen on the Bass Harbor tour.

CONSERVATION CONCERNS

Acadia is located within a day's drive of some of the largest cities in eastern America. When 2.4 million summer people (as the locals call visitors) arrive, traffic snarls and difficulty in finding parking become commonplace.

Most of Mount Desert Island is served by a free shuttle bus that connects campgrounds, resorts, inns, and Bar Harbor and other towns to the park's principal attractions including the most popular trails. Shuttles run at least once per hour on seven shuttle routes.

Acadia is located downwind of some of the largest cities on the East Coast. Periodically, high levels of pollutants occur at Acadia. These include ozone, sulfur dioxide, nitrogen oxides, particulates, and mercury. In 2001, there were ten days when air was unhealthy in the park due to ozone. Pollutants are being monitored, and research is under way to determine the effects of air pollution on the park's terrestrial and aquatic ecosystems.

The method of Acadia's establishment resulted in a patchwork of park and private ownership. Sewage disposal, landfills, oil spills, and other activities on private land may result in pollutants entering park waters.

Deposition of pollutants in lakes and ponds from the atmosphere is also occurring. The pH of precipitation at Acadia measures between 4.4 and 4.6. One fog was measured with a pH of 3.0, the equivalent of grapefruit juice. Measurements of mercury concentrations in freshwater fish are among the highest in the United States. Monitoring and research continue to help determine what is happening to water quality at Acadia.

Acadia is threatened by exotics, both plant and animal. Twelve species of exotic plants are of major concern, including garlic mustard, Japanese barberry, and purple loosestrife. The gypsy moth has spread to Acadia, playing havoc with the park's forests. Park waters have been invaded by the zebra mussel. Ridding the park of any of these will be extremely difficult, expensive, and time-consuming.

ARCHES
A Place of Nature and Time

Moab, Utah
ACRES: 76,519

J ohn Wesley Wolfe was a big man. The Civil War artilleryman had fought through much of Grant's campaign in the Vicksburg area and had fortunately avoided Confederate shot and shell. But during the siege of Vicksburg, he severely injured his leg while trying to move a cannon out of the mud. Wolfe was given a disability discharge, but for him the war was over.

Though he would walk with a cane for the rest of his life, Wolfe had a severe case of wanderlust. In the company of his son Fred, Wolfe left his wife and three other children behind in Ohio and sought his fate in the west. The better, well-watered lands had already

Photographs on preceding pages:

(Pages xii–1)
A black-tailed prairie dog appears near a bison herd in Wind Cave National Park.

(Pages 2–3)
Gray wolves stalk an elk herd in Yellowstone National Park.

(Pages 4–5)
Tourists watch a distant pack of wild wolves in Yellowstone National Park.

(Pages 6–7)
An American bison after a winter snowfall in Yellowstone National Park.

The namesake of the park, stone arches provide a window onto a spectacular scene. Arches National Park contains two thousand arches, more per area than any other place in the world.

been homesteaded, so Wolfe's search took him to the desert country of eastern Utah. There he found a place near the Colorado River along Salt Wash with enough water and grass to make a living. The two men built a cabin and began a small cattle operation. Summers were hot and the winters surprisingly cold. The ranch was lonely and they never struck it rich, but Wolfe and his son learned to love their new home and remained there for twenty years.

Visit Arches National Park today and you'll find the Wolfe Ranch is the centerpiece of the park's cultural history. Would you have chosen such a place to live? Could you have withstood the remoteness and the hot, dry climate? Do you think Wolfe could have ever foreseen the time when three quarters of a million people would visit this park to enjoy the marvelous scenery?

Arches National Park was designated a national monument in 1929 and received national park status in 1971. As the name suggests, the park is noted for having the greatest density of arches in the world. An arch is created by the action of wind and water, as opposed to a natural bridge, which is the result of a flowing stream. More than two thousand arches have been cataloged, ranging from three-foot openings (the smallest officially to be considered an arch) to the 306-foot Landscape Arch. New arches continue to be created today. Most arches are contained within the salmon-colored Entrada Sandstone, deposited during the Jurassic period 150 million years ago.

The arches are only part of the story. The park contains seventy-six thousand acres of some of America's best scenery. The eighteen-mile Arches Scenic Drive provides access to concentrations of arches at Devil's Garden and The Windows. Delicate Arch, probably the most photographed arch in the

area, is accessible by trail from the Wolfe Ranch parking area. Another popular feature is the Big Balanced Rock, 128 feet high including pedestal and weighing more than thirty-five hundred tons.

Arches National Park is a "cool" desert environment. Annual rainfall is about ten inches, much of it falling during summer thunderstorms. Summers are typically hot and dry with temperatures sometimes exceeding 100°F. Winters can be cold, and occasionally temperatures drop below zero. Elevation ranges from about 4,085 feet at the Visitor Center to 5,653 feet on Elephant Butte. The similarity of climate throughout the park and the narrow differences in elevation result in limited biological diversity. Even so, there are more than fifty species of mammals found here.

The desert environment requires animals to adapt for survival. Most of the park's mammals are small rodents. The white-tailed antelope squirrel and the Colorado chipmunk are regularly seen during daylight. The only other squirrel in the park is the rock squirrel. Arches' woodrats, or "pack rats" include the desert woodrat, the Mexican woodrat, and the bushy-tailed woodrat.

The largest mammal likely to be seen is the mule deer. They are found throughout the park and are active early and late in the day. Wildlife managers have been successful in reversing the decline of the desert bighorn sheep though they are still rare.

There are two species of lagomorphs in Arches, the desert cottontail and the black-tailed jackrabbit. Both are quite abundant and can best be seen along highways at dusk and dawn.

Among the smaller predators, the coyote is quite common but is most active at night. The gray fox and kit fox are common, but both are nocturnal and seldom seen. All the mustelids are rare except the uncommon badger. These include the river otter, western spotted skunk, and striped skunk.

Both the bobcat and the mountain lion are considered rare; however, tracks of both have been seen in the area of Delicate Arch. Whether they are resident here or transient is not known.

CONSERVATION CONCERNS

The area around the park is characterized by tremendous growth in tourism. The Moab area has become a mecca for mountain bikers, river floaters and rafters, campers, and hikers. Large numbers of visitors can have an adverse impact on the natural environments of the park. For example, an important part of the desert ecosystem is cryptobiotic soil crust, which is important in erosion control, water retention, and providing habitat. Trampling by humans can be devastating.

BADLANDS

The Wilderness of Pinnacles and Spires

Southwestern South Dakota

ACRES: 244,000

*L*es *mauvaises terres à traverser.* French-Canadian fur trappers came to this part of present-day South Dakota in the early eighteenth century, apparently leaving without pleasant memories—"bad lands to travel across." The Dakota (Sioux) had a similar name for the area—*Mako* (land) *Sica* (bad), though it is thought they frequently used the area. Today we see nothing "bad" about this place; instead, it is the destination of nearly nine hundred thousand travelers annually, coming to contemplate nature's artwork of wind and water.

The badlands story begins about seventy-five million years ago when this region was covered by a shallow sea. Over the course of time, the remains of millions of sea creatures collected on the sea bottom to form the Pierre shale. This pink and yellow formation today contains evidence of ancient fish, reptiles, turtles, and birds.

Much later, plate tectonics caused the sea bed to rise, ultimately creating dry land. Initially the landscape was a warm, humid forest, but gradually the climate became drier and cooler, forming first savannahs and later grasslands. The mammals of the Oligocene period (twenty-three to thirty-five million years ago) were a varied lot. They included *Mesohippus* (Oligocene horse), *Hyracodon* (Oligocene rhinoceros), and *Hoplophoneus,* the saber-toothed cat. Three interpretive trails are provided to help visitors understand rocks and fossils of the prehistoric badlands.

But consider modern-day mammals. The paved Badlands Loop Road between the Northeast Entrance and the Pinnacles Entrance offers good viewing opportunities. Both entrances are within ten miles of Interstate 90. North of the Badlands Loop Road is primarily grassy prairie while the badlands are to the south. More adventurous visitors may wish to visit the Sage Creek area, accessible by the unpaved Sage Creek Rim Road.

Watching the antics of the black-tailed prairie dog is a popular pastime. The best place to see them is the Roberts Prairie Dog Town. Don't feed them—they bite!

Numerous other rodents are common throughout the park. The least chipmunk is probably most numerous. The thirteen-lined ground squirrel may also be spotted in the prairie dog towns. Porcupines can sometimes be found in the thickets of Sage Creek Wilderness. Beaver and muskrat may be encountered wherever there is permanent water. The bushy-tailed woodrat is common in the badlands of the park.

Other small mammals are present. Perhaps the most frequently seen in the park is the desert cottontail. Other lagomorphs include the eastern cottontail, the white-tailed jackrabbit, and the black-tailed jackrabbit.

There are several species of herd animals in the park. The most common is the mule deer, which can usually be viewed on any drive through the park. There are white-tailed deer here too, though they are much less common and are usually found in wooded thickets.

Another common herd animal you're likely to see is the pronghorn. Pronghorns are common along the Badlands Loop Road especially to the north in grassy areas.

As was true of most places in the West, the buffalo, or officially, the bison, had been extirpated from the Badlands during the late nineteenth century. They were successfully reintroduced in 1963 here in the Sage Creek Wilderness. Today, several hundred bison live in the sixty-four thousand-acre wilderness. They are not, however, free roaming due to adjacent ranchers' concern of the possibility of spreading brucellosis to their cattle. You may see bison along the Sage Creek Rim Road and even perhaps in the Sage Creek Campground.

Like the bison, bighorn sheep were eradicated from the park; the last known individual was shot in the South Unit in 1926. And like the bison, they were successfully reintroduced in 1964. Unlike the bison and the pronghorn, bighorn are much more at home in the badlands than in the prairie.

Early French explorers called this place les mauvaises terres à traverser, *which translates loosely as "bad lands to travel across." Today, paved roads and surfaced trails make access easy for travelers, yet retain a sense of what the area must have once felt like.*

Among the carnivores, most common is the coyote. They may be seen throughout the park, especially near prairie dog towns. The rare swift fox is a prairie dweller, especially around prairie dog towns. The mountain lion and the bobcat are also present in prairie and badlands but neither is often seen. Three members of the weasel family reside in Badlands: the badger, the striped skunk, and the eastern spotted skunk.

Badlands is the site of hope for a member of the weasel family. In the late 1970s, most biologists believed that the black-footed ferret was extinct. It had been considered one of the most endangered land mammals in North America since its prey consisted entirely of prairie dogs, a species whose habitat was being tremendously reduced nationwide. But in 1981, a population of one hundred ferrets was found in Wyoming. Though canine distemper killed all but eighteen of them, those remaining were placed in captivity and used for breeding for reintroductions. Badlands National Park was selected as a site for reintroduction, and in 1994, several ferrets were carried into Sage Creek Wilderness and released in prairie dog towns. They continue to hold their own today, a symbol of America's wildlife heritage.

CONSERVATION CONCERNS

Though Badlands National Park is protected by statute, problems still exist. Exotic plants have become a severe threat, which could cause serious alterations in habitats if left unchecked. The Park has created the Badlands Weed Management Area to control these pests using chemicals, prescribed burning, mowing, and biological control. Pollutants borne by wind and water are also potential problems, and current legislation is not fully protecting park resources.

Big Bend National Park, at a major curve in the Rio Grande, is a great example of the vegetation and geology of the Chihuahuan Desert. A combination of weather conditions and excessive water withdrawal for irrigation sometimes causes the river to run dry.

BIG BEND
Where the Border River Bends

Big Bend National Park, Texas
ACRES: 801,163

People have many differing views about deserts. Some think they're all Sahara-like sand dunes with widely scattered oases the only relief from wind, sand, and sun. Others have cartoon-like images of coyotes chasing roadrunners through red-rock canyons beneath thick-armed saguaro cacti. It's always hot, it's always dry, and everything has either a thorn or a bite. But there are no dunes here, nor are there any saguaros, though there are coyotes and roadrunners! This is the Chihuahuan Desert, and Big Bend National Park is one of the best places in America to get to know it up close.

The major land form here is the Chisos Mountains. The elevation difference between the Rio Grande and Emory Peak is nearly six thousand feet, resulting in tremendous diversity. Close to the river is desert shrub with ocotillo, creosote bush, and prickly pear cactus abundant. By the time you reach the Basin, grasses predominate. If you take a hike higher, you'll find pinyon pine and juniper, and in two canyons, ponderosa pine. In addition, the southern border of the park is the Rio Grande, home to riparian-dependent species; Mexico is beyond. The amount of wildlife here is exceptional. There are more species of birds in Big Bend than in any other American national park. The park's official mammal list includes seventy-five species.

Lagomorphs are common at Big Bend. They include the desert cottontail, eastern cottontail, and the black-tailed jackrabbit.

Hiking in the high country, you may see the rock squirrel. Lower, the most common squirrel is the Texas antelope squirrel; the spotted ground squirrel is uncommon. Other rodents include the woodrat or pack rat. There are three species in Big Bend, the most common of which is the white-throated woodrat. The others are the Mexican woodrat and southern plains woodrat.

A mammal you'll see only along the Rio Grande is the beaver. An exotic animal, sometimes mistaken for a beaver, is the nutria. There are populations of them further downstream; they have been reported in Big Bend at times.

Big Bend's most common predator is the gray fox. Coyotes are commonly encountered. Along the river, the raccoon is common. The ringtail is secretive and seldom seen. The only common member of the weasel family to be found at Big Bend is the striped skunk, a real problem in campgrounds.

The story of the black bear at Big Bend is especially noteworthy. The original bear population in the area was destroyed by hunting by the 1930s. For the next half century, the only bears in the park were wandering individuals from Mexico that did not take up residence. In the 1980s, however, some moved in permanently from the Sierra del Carmen in Mexico, and today the Chisos supports a resident population of about twenty adults. Natural repopulation of a species doesn't happen often, so the bears' return is a special event.

Another large predator here is the mountain lion. Encounters between humans and lions have increased throughout the lion's range as more people enter its territory. There are between twenty-five and thirty lions in the park, and annual sightings average about 150 per year. In the last twenty years, there have been three attacks on humans by lions in the park but no fatalities. Aggressive lions were subsequently destroyed. Lions play a critical role in the park's ecosystem and their loss would be catastrophic. Consequently, park staff is working on a plan to prevent encounters in order to allay visitor fears and to protect the lion. This includes temporary closure of some sites due to lion activity.

You're likely to see one of the ungulate species. The mule deer is a common resident in the desert areas. The other species, the Carmen Mountains white-tailed deer, lives in the Chisos Mountains. The javelina, or collared peccary, is common at the park's lower elevations. They roam in groups of up to twenty and live in a variety of habitats.

CONSERVATION CONCERNS

Like many parks, Big Bend faces problems with air and water pollution, but its location on the Mexican border makes the situation more serious. Much of the water from the upper Rio Grande is used by water interests in the United States. The river actually goes dry below El Paso at times. Most of the water passing Big Bend comes from the Rio Conchos, which originates in the Sierra Madre in Mexico. Dealing with increased water demand and possible water quality degradation is a major concern of park management for the foreseeable future.

During the summer, you can usually see between thirty and ninety miles. About 6 percent of the time, visibility is reduced, some days to less than ten miles. What is the cause? About half comes from coal-fired power plants on both sides of the border. Weather patterns also bring pollutants, including sulfates and nitrates, to Big Bend from the Mexico City area. Continued population growth will only make matters more difficult.

As we have seen, many mammals ignore international boundaries. Migration between countries happens frequently. We need to continue to work with Mexican authorities to protect animals throughout their life cycle.

BISCAYNE
The Park Beneath the Sea

Homestead, Florida

ACRES: 172,924

You probably have not come to Biscayne to look for mammals. The ocean drew you here, and you probably can't wait to swim, fish, snorkel, or take the concessionaire boat to camp on Boca Chita or Elliott Key. Biscayne National Park is mostly water; 95 percent of its 173,000 acres to be exact. Even so, there are some interesting animals to be found here.

Most of the half million visitors come to Biscayne in private boats. There is mainland access at the Dante Fascell Visitor Center in Homestead. The center, opened in 1997, replaced an earlier visitor center destroyed by Hurricane Andrew. Concessionaire watercraft give those without their own boats a way to get onto or under the water.

The park contains the northernmost coral reef in the United States. You'll see the chain of forty-two tropical isles that make up most of the land portion of the park. The largest, Elliott Key, is considered the northern beginning of the Florida Keys. These islands were created by coral, while the islands to the north of Elliott Key are a combination of sand and the action of coral.

Most of the animals you'll see on the islands are insects, but there are mammals too. The raccoon (especially around campgrounds), the marsh rabbit, the

The reefs of Biscayne National Park are home to many sea creatures, including several species of coral. The fragile beauty of the park's ecosystem is its main attraction. Unfortunately, a variety of factors in recent years have resulted in Biscayne being listed among the most endangered parks.

cotton mouse, and the Key Largo woodrat (a federally endangered subspecies of the eastern woodrat) live here. On Elliott Key you might even see a Mexican red-bellied squirrel. How did it get here? Apparently, a local resident introduced them from northern Mexico because he liked them.

According to the park's mammal list, several marine mammals are listed as present in the park. Whales considered present in Biscayne waters are the finback, humpback, northern right, sei, and sperm whales. Another marine mammal present at Biscayne is the Atlantic bottlenose dolphin.

There is one endangered mammal in the park, one with considerable charisma, the manatee. Protected now by the Marine Mammal Protection Act, the manatee was almost driven to extinction by hunting. There are about two thousand manatees in south Florida. They live in Florida's coastal waters in winter but migrate north and west during summer months. The herbivorous giant may grow to more than fifteen hundred pounds. Unfortunately, the manatee's gentle nature can cause problems. Most manatees bear scars on their backs or tails from collisions with the propellers of outboard motors. Many of these encounters are fatal. Habitat destruction and water pollution are also threats. Continued vigilance is necessary to protect this fascinating animal.

CONSERVATION CONCERNS

Each year the National Parks Conservation Association (NPCA) releases a list of America's "Ten Most Endangered Parks." Unfortunately, in 2004, Biscayne National Park is on the list. There are several reasons. According to NPCA, overfishing is causing the decline of several fish populations. Pollution and unthinking boaters and divers are causing harm to the coral.

The northern edge of the park is less than twenty miles from Miami, a metropolis of more than 2.5 million people. As population increases, housing developments will spread, resulting in the impeding of fresh water into the park's ecosystem.

The plight of the manatee is also a concern. Good citizenship is the best response to these problems. Review park management documents on the park's website. Tell Congress and park staff what you think. Your voice does count.

BLACK CANYON OF THE GUNNISON
Sheer Walls and Narrow Passages

Montrose, Colorado

ACRES: 30,244

As the Gunnison River makes its way through the Black Canyon of the Gunnison, the loss in elevation in just forty-eight miles exceeds that of the entire length of the Mississippi River. The result is a canyon so deep and narrow that sunlight rarely strikes the bottom, hence the canyon's name. A paved highway connects several viewpoints along the south rim. At the end of the road is Warner Point Nature Trail where the canyon is deepest, 2,772 feet. If you're fit, there are trails to the river but they are extremely steep. The elevation difference between the rim and the river is nearly two thousand feet.

Most mammals at Black Canyon of the Gunnison are nocturnal and seldom seen. The ones you will see depend on which park habitat you're visiting. The western portion of the South Rim exceeds eight thousand feet in elevation. The forest at this elevation is primarily pinyon pine and junipers, and since the trees are shorter here than elsewhere in the park, it is referred to as the "pygmy forest." Short does not mean young; some pinyon pines near High Point have been found to be more than seven hundred years old. Some of these ancients are beginning to die. One answer perhaps is the presence of the porcupine. You may not see one, but you will see signs of their feeding on pinyon bark. More study is needed to fully understand the relationship between the porcupine and its habitat.

The eastern part of the South Rim consists of Oak Flats, the primary tree being the Gambel oak. The oaks form a dense thicket providing good cover and an abundance of acorns. Many mammals are dependent on the oak for food and shelter. These include the Colorado chipmunk, least chipmunk, rock squirrel, and golden-mantled ground squirrel. In more open areas, you might see the yellow-bellied marmot catching some sun. Coyotes, bobcats, and gray foxes are the primary predators.

Some of the larger mammals are found here as well. Year round at dusk and dawn, mule deer browse in the oaks, often along the roads. During the winter, elk may migrate into these oak woodlands for shelter from the harsh winter. They may remain until June. Oddly, eating too much Gambel oak is toxic to cattle.

Mountain lions sometimes inhabit the park. Their chief prey is the mule deer and smaller mammals. Bighorn sheep can sometimes be seen inside the canyon. Their adaptations to the steep rocky habitat allow them to avoid predators like the mountain lion.

Some of the members of the weasel family in the park include the long-tailed weasel, badger, and striped skunk. The lagomorphs in the park are Nuttall's cottontail and the white-tailed jackrabbit.

Along the river is a riparian forest consisting of several tree species including boxelder and cottonwood. Mammals that may be encountered after a hike to the river are the beaver, raccoon, and mule deer. If you camp at the bottom, you might see a ringtail.

CONSERVATION CONCERNS

Black Canyon of the Gunnison is a relatively small park containing only 30,045 acres. Its location somewhat off the beaten path in western Colorado results in a modest visitation of less than two hundred thousand annually. Even so, development nearby is partially responsible for the loss of habitat for much of the wildlife found here. This is especially true for some of the larger mammals. But the presence of an awe-inspiring canyon, a very interesting human history, and the chance to enjoy some of the west's best landscapes make a trip to Black Canyon of the Gunnison well worthwhile.

BRYCE CANYON

Sculptured Rocks and Whimsical Formations

Bryce Canyon, Utah
ACRES: 35,835

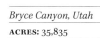

"A hell of a place to lose a cow." So said Ebenezer Bryce, namesake of Bryce Canyon, one of America's most beautiful national parks. Bryce Canyon isn't really a canyon; it's the Paunsaugunt Plateau, its eastern edge eroding into colorful and weirdly shaped hoodoos, or

natural rock columns. Enjoying the formations from above or hiking a trail below the rim is the primary reason people come to Bryce.

The park's 35,835 acres represent wide altitude extremes. At the southern end of the scenic drive is Rainbow Point, 9,115 feet above sea level. Below the rim, elevations fall away to less than seven thousand feet near the east edge of the park. In between are three main ecosystems—the pinyon-juniper type below the rim, the ponderosa pine woodlands on northern parts of the plateau, and on the higher southern part of the plateau, Douglas fir, white fir, and blue spruce are found.

The widest variety of mammals in the park is the rodents. The common ones are the least and Uinta chipmunk and the rock squirrel. An abundant rodent throughout the park is the golden-mantled ground squirrel. Unfortunately, many visitors have been feeding these cute critters.

The bushy-tailed woodrat and the desert woodrat are resident in the park. Beaver and muskrat are present but uncommon. The red squirrel is common in spruce-fir forests. Porcupines can be encountered anywhere there are conifers. The yellow-bellied marmot may be seen sunning itself.

Prairie dogs used to be one of the commonest mammals in the west. The westernmost of the five prairie dog species, the Utah prairie dog, lives in Bryce Canyon. Over the years its population dwindled, and it was feared that extinction was imminent. Finally in 1973, the Utah prairie dog was listed as an endangered species. Today, there are two hundred of them in "towns" in Bryce Canyon, the nation's largest protected population. The species has been downlisted to threatened, but its numbers remain low.

More than one-half mile deep, the Black Canyon of the Gunnison is one of western Colorado's deepest geological sites and certainly one of the most spectacular sites in the American West. On rare occasions a mountain lion can be seen in the park, hunting for mule deer and other mammals.

The common black-tailed jackrabbit can be found throughout the park while the mountain cottontail lives in the higher elevation forests. The desert cottontail is found in sagebrush and pinyon-juniper forest.

The predators are present at Bryce but are rarely sighted. The mountain lion and bobcat live throughout the park. The black bear is widespread in the park but is rare. Among the smaller predators, the coyote is the most common. The nocturnal gray fox is sometimes seen along park roads at night. The weasel family is well represented and widespread at Bryce. It includes the western striped skunk, spotted skunk, badger, and long-tailed weasel.

Among the ungulates, the abundant mule deer seems to be everywhere. They are common where people are, in campgrounds and at rim overlooks. Another ungulate in Bryce Canyon is the elk, but they are present only in the winter. Pronghorns are common in the region but their presence is variable.

CONSERVATION CONCERNS

Perhaps the biggest concern at Bryce Canyon is air pollution. The parks of the Colorado Plateau are known for their stunning red-rock scenery. But as air quality declines, visibility decreases. The park installed visibility monitoring equipment at Rainbow Point focused south toward Arizona and New Mexico. Being able to see points in either state has become much more infrequent.

South of the park is the Kaiparowits Plateau, underlain by millions of tons of coal reserves. Mining these reserves is not in the offing today, but as our nation's concern for energy becomes more acute, will we sacrifice Bryce Canyon's scenery in our search for a viable energy policy?

CANYONLANDS
Colors of the Primitive Desert

Moab, Utah

ACRES: 337,598

" Island in the Sky," "The Maze," "The Needles," "Paul Bunyan's Potty." Interesting names all, and all place names in Canyonlands National Park. One of the newer parks on the Colorado Plateau, Canyonlands was created in 1964. Additions were made to the park in 1971, expanding it to 337,598 acres. Canyonlands is one the wildest areas in the National Park System outside Alaska.

Like other Utah parks, Canyonlands is desert red-rock country. Erosion has created fantastic rock formations with stunning overlooks. Two highways provide access to the park. In the Island in the Sky, a drive to the end of the scenic highway at Grandview Point brings you to one of the great sights in the

American west. Another drive in the southeastern part of the park allows access to The Needles. Here, instead of vast panoramas, you are among the formations, getting a much different perspective.

Canyonlands has something most other parks don't have—a system of primitive roads suitable for four-wheel-drive vehicles. These primitive roads provide a unique experience for many visitors and include camping opportunities at sites on the White Rim and in The Maze. Most of the park, however, is accessible only by foot.

Two of the west's major rivers meet in Canyonlands. The Green and Colorado Rivers come together at the Confluence, creating an even greater Colorado. Much of the river mileage provides world-class rafting opportunities including the Colorado River's famous Cataract Canyon. Both rivers exhibit significant riparian habitats, increasing the kinds and amount of wildlife typically found in the desert. There are nearly fifty mammal species in Canyonlands. Many of the animals are small desert dwellers; others are adapted to higher or wetter habitats.

Small mammals abound in Canyonlands. Common squirrels at Canyonlands include the Colorado chipmunk, the white-tailed antelope squirrel, and the rock squirrel. There are three woodrats in Canyonlands including the common desert woodrat and the uncommon Mexican woodrat and bushy-tailed woodrat. The porcupine is found wherever there are pinyon pines, while beaver are found primarily in the riparian zones. The desert cottontail and the black-tailed jackrabbit abound.

Numerous predators prey on the smaller mammals. The coyote and kit fox are common. Along the rivers, you're more likely to see the gray fox. Both the bobcat and the mountain lion inhabit Canyonlands but only an extremely fortunate visitor will see either one. The black bear may be an occasional transient in Canyonlands.

The largest mammal you're likely to see is an ungulate, the mule deer. They are common, and can be found in all areas of the park, from the plateau to the rivers.

Historically, the desert bighorn were quite abundant throughout the southwestern deserts, but due to hunting and transmission of diseases from domestic sheep their populations became severely reduced. At the time of establishment of Canyonlands, about a hundred bighorns inhabited the park. With park protection and the removal of grazing permits within the park, bighorn populations grew. Growth was sufficient to transplant sheep to nearby parks and to spatially expand the bighorn's presence in Canyonlands. Today there are about 350 bighorns in the park.

CONSERVATION CONCERNS

As with other parks on the Colorado Plateau, a haze often clouds the view. This haze comes primarily from coal-fired power plants and other industries, carried long distances by wind currents. Even though Canyonlands receives the highest protection under the Clean Air Act, air pollution still occurs.

Exotic plants and animals are a continuing problem. Along the rivers, tamarisk, an Asian import, crowds out native vegetation along the riverbanks. Wildlife dependent on native plants can be severely harmed. Other exotic plants are found elsewhere in the park. It is estimated that forty species of non-native fish live in the Colorado River. In fact, the most frequently seen fish are the carp and the channel catfish, both exotics.

Canyonlands is a tremendous resource. But the park, according to the National Parks Conservation Association (NPCA), is "incomplete and vulnerable." Land uses outside park boundaries pose real threats, perhaps the most worrisome being the potential for oil and gas development east of the park. NPCA's solution is to expand the park's acreage. Citizens should make their elected officials aware of their views on the allocation of public lands to parks and other uses.

Wind and water have sculpted the multi-hued rocks of Canyonlands National Park into extraordinary shapes that provide a stunning vista. The park is home to about 350 bighorn sheep.

CAPITOL REEF
Land of the Sleeping Rainbow

Torrey, Utah
ACRES: 241,904

Capitol Reef—the name suggests a nearby sea, but it actually comes from a geologic formation called Waterpocket Fold, created when sedimentary deposits were warped by an extensive mountain-building event. The north-south trending fold acts as a "reef" for travelers moving east or west. There were passages through the reef;

The sheer cliffs of Capitol Reef presented travelers in days past with a seemingly insurmountable impediment as they moved across what is now Utah.

early travelers used Grand Wash, even driving their Model Ts through the unpaved wash. Rock messages from these early travelers are common along its walls.

Another passage was the valley of the Fremont River, the primary way into the area more recently. When Capitol Reef was originally created as a national monument in 1937, the road along the river was a dirt track. The highway was paved in 1962, making Capitol Reef accessible to a coming boom in tourism. Capitol Reef achieved national park status in 1971. Visitation in 2003 exceeded a half million visitors.

Where does the name "Capitol" come from? There are many interesting rock formations in Capitol Reef that attract visitors to the park. One huge Navajo sandstone dome, Capitol Dome, apparently looked to early settlers quite similar to the dome of the U.S. Capitol Building in Washington.

Settlers came to Capitol Reef in 1880, founding the village of Fruita. Never large, Fruita's last resident moved away in 1969. Even though this is a desert getting only seven inches of rain annually, the Fremont River offered opportunities for irrigated agriculture. There is a legacy of almost three thousand acres of orchards containing numerous kinds of fruits left behind. You can pick them in season for a fee. The area around Fruita, well watered by the Fremont River, is the prime viewing

area for the mammals of the park. Most park facilities are located nearby.

The mammal you're most likely to see is the mule deer. They inhabit most habitats in all parts of the park. Other large herbivores are very rarely seen. The American bison lived here historically, but only a few introduced individuals enter the far eastern part of the park today.

Water attracts wildlife, and the Fremont River, especially around Fruita, provides habitat for many mammals. Beaver inhabit Halls Creek as well as the Fremont River. The muskrat also calls the Fremont home but there are fewer of them. Mink inhabit the park's riverbank areas. The striped and western spotted skunks are never far from water. Finally, the ringtail is a native of rocky areas close to rivers. Badgers inhabit the area along the Fremont River at the east boundary of the park.

The white-tailed antelope squirrel is abundant in lower areas. In rocky canyons near the river lives the rock squirrel. In lower desert areas of the North District, the least chipmunk is found. Eradicated through poisoning, the Utah prairie dog was reintroduced at Capitol Reef; a small population lives in the far North District. The desert woodrat is found in desert areas especially on rocky slopes.

There are rodents in the higher areas of the park as well. Yellow-bellied marmots are typically found in higher elevations, but here they are found along the Fremont in the Fruita area. You may also find the Colorado chipmunk. The porcupine is common in forested areas but may also be found in brushy areas.

There are predators in Capitol Reef. The most abundant predator there is the gray fox. The coyote is found in the Fruita area but also elsewhere. In the North District, the red fox competes with the coyote. The kit fox may be expanding its range into the park.

CONSERVATION CONCERNS

Air pollution is a concern at Capitol Reef as it is with other Utah parks. The culprits are coal-fired power plants and other industries. Capitol Reef has been the site of an ongoing political controversy. The issue concerns ownership of some of the roads in the southern and eastern parts of the park, notably the Burr Trail. Local governments want ownership and the right to pave these roads for better access and increased tourism. The National Park Service prefers to maintain the roads unpaved to reduce visitation and to enhance wildlife numbers. Years may pass before this controversy is settled.

CARLSBAD CAVERNS

America's Deepest Limestone Cave

Carlsbad, New Mexico

ACRES: 46,766

On the northern fringes of the Chihuahuan Desert, you'll find an interesting geological feature—the remains of a four hundred-mile-long fossil barrier reef, formed under a Permian-age shallow sea. The southern part of this reef is Guadalupe Mountains National Park in Texas. To the north, it's what is inside the reef that interests people—Carlsbad Caverns. With some of America's most spectacular cave scenery, Carlsbad Caverns was first designated a national monument in 1923 and received national park status in 1930. The park contains 46,766 acres, three-fourths of which have been designated as wilderness.

Almost all of the visitors to Carlsbad Caverns (460,000 in 2003) are people who enter the cave. Of course, most of the sixty-four species of mammals in the park are on the surface, but visitors frequently encounter the park's underground mammals, the sixteen species of bats.

By far the most important bat species is the Brazilian free-tailed bat. This is the mammal of the world-famous bat flight. Approximately 350,000 free-tailed bats roost in the cave where they give birth and nurse their young. Roost sites are not on cave tour routes. The bats are only at Carlsbad Caverns seasonally, migrating to Mexico and Central America for the winter, due to a lack of their primary food—insects—at Carlsbad Caverns during the colder months.

Just before sunset each summer night, visitors gather at the small amphitheater at the cave entrance for the nightly bat flight. After a short presentation by park rangers, visitors sit back and wait. At first, a few scattered individuals appear. Gradually, the volume of bats increases until the sky looks as though filled with spiraling smoke. Once they reach the proper altitude, the bats disperse across the landscape for a night of feasting on their insect prey, returning just before dawn. Though the volume of bats returning is less concentrated, some of the returning bats make spectacular dives into the cave mouth.

Some visitors may notice other bat species as well. The common species are the fringed myotis, the California myotis, the Western pipistrelle, and Townsend's big-eared bat. Like the free-tailed bat, these bats are seasonal residents, migrating south for the winter.

A few surface mammals, especially predators, spend time around the mouth of the cave. Three species of skunk fill this niche: the striped skunk, western spotted skunk, and western hog-nosed skunk. Walking the Desert Nature Trail near the cave mouth, you might also see a white-throated woodrat, ringtail, long-tailed weasel, or raccoon. Coyotes are often around the entrance, but are common elsewhere.

The surface of Carlsbad Caverns National Park is desert, ranging in elevation from the low desert at thirty-six hundred feet to the highest elevation at nearly sixty-four hundred feet. You may see surface mammals along the five-mile-long access road that brings you to the visitor center where cave tours begin. The most abundant mammal is the mule deer. Lagomorphs include the desert cottontail and black-tailed jackrabbit. Porcupines are common and most often seen in stands of trees.

Desert parks have plenty of rodents. In the higher parts of the park, you may find the rock squirrel and the white-throated woodrat. Lower, you may see the Texas antelope squirrel and the southern plains woodrat, or pack rat.

The park does have some larger predators. The mountain lion lives here. A 1980s lion survey radio-collared twenty-two lions within the park and from the data estimated a population of fifty-eight lions. Lion signs are assessed annually; results indicate that the number of lions in the park is remaining stable.

CONSERVATION CONCERNS

The biggest concern at Carlsbad Caverns is the bats. Since they feed primarily outside the park, there are no assurances that agricultural pesticides will not harm them. Perhaps a bigger concern is what happens to them in their wintering grounds in Mexico and Central America. Migrating wildlife frequently find their habitats destroyed by development. Park managers at Carlsbad Caverns have no recourse when this happens. Education about the value of bats, especially as a huge consumer of insects, is necessary here in the United States and in countries where wintering bats make their homes.

Nearly one-half million visitors see the colorful cave formations of Carlsbad Caverns each year. The deepest of America's limestone caves, Carlsbad has a huge colony of Brazilian free-tailed bats, numbering approximately 350,000.

CHANNEL ISLANDS

Pearls in the Pacific

Visiting one of the eight Channel Islands involves a trip by boat from nearby Santa Barbara. The park is an especially good place to view marine mammals, with thirty-five species recorded in nearby waters.

Ventura, California

ACRES: 249,561

America is an automobile country. We travel by car and most vacations include long stretches of highway. But some national parks don't lend themselves to car travel; in fact, there are a few where your car won't do you much good. Channel Islands, off the coast of southern California, is one such park.

There are eight Channel Islands, five of which make up Channel Islands National Park: Anacapa, San Miguel, Santa Barbara, Santa Rosa, and Santa Cruz. The total acreage of the park is about a quarter

of a million, almost equally divided by land and water. Though you aren't far from the highly populated Los Angeles area, the islands have remained isolated. No motor vehicles mar the islands' landscape.

Access to the Channel Islands is mostly by boat. A park concessionaire operates out of Santa Barbara harbor with day trips or longer multi-day trips available. It is also possible to fly to Santa Rosa Island from the airport at Camarillo, California.

There are numerous activities available on the islands. Many people go for the day to picnic, scuba dive, snorkel, view wildlife, fish, and hike. Primitive camping is also available. Several interpretive programs are offered including guided walks and exhibits.

Channel Islands National Park is an internationally significant natural resource hosting more than two thousand species of plants and animals. There are several different plant ecosystems on the islands ranging from coastal dunes to woodlands of oak and other hardwoods as well as conifers such as the Torrey pine. The islands are characterized by a high degree of endemism. Marine mammals are perhaps the most exciting wildlife to be found here. There are also a few terrestrial species and, unfortunately, feral species as well.

The park lists thirty-five species of marine mammals that have been observed in its surrounding waters. Several whale species migrate through the park. The gray whale migrates between Alaska and Mexico's Baja California and uses the park as a resting place on both segments of its journey. They may be present from December through May. The humpback whale is a summer visitor and frequently can be seen around San Miguel and Santa Rosa Islands. The blue whale, the world's largest mammal, also appears in summer and may be seen in the company of humpback whales.

Killer whales or orcas inhabit Channel Islands. They seem to be most common during the gray whale migration when they prey on gray whales as well as the park's resident seals and sea lions (pinnipeds). The large pinniped population in the park attracts orcas all year and a few of them may now be residents.

Smaller cetaceans are present; most likely to be seen is the common dolphin. Most boat trips to the islands will encounter several. Their total population in the park is estimated to be near fifteen thousand. Other resident dolphins and porpoises include the bottlenose dolphin and Dall's porpoise.

One would expect to find sea otters among the islands; a hundred years ago you would have. But hunting caused the local extirpation of sea otters at Channel Islands. A few have returned, but it will be a while before tourists are likely to see one.

Channel Islands is home to several species of pinnipeds that come here to breed and give birth. As many as 150,000 California sea lions live around the islands; their population continues to increase. Resident on San Miguel is the northern fur seal, which visits each May to breed and give birth. The northern elephant seal comes to the islands twice a year, in June to molt and in January to breed. After being hunted nearly to extinction, their numbers continue to grow. Finally, the harbor seal is a year-round resident of the islands, though its population has declined to about twelve hundred.

The largest native terrestrial mammal is the island fox. About the size of a housecat, the fox is a descendant of the mainland's gray fox. In 1993 the National Park Service initiated a monitoring program to assess fox populations. In 1994, about four hundred foxes were found on San Miguel, but five years later only a few dozen could be found. A similar problem was found on Santa Rosa and Santa Cruz Islands. One cause is the presence of canine diseases such as heartworm and parvovirus.

Another cause is predation from the golden eagle. No golden eagle nest had ever been found on the islands until 1999. The raptor had apparently become resident, preying upon exotic animals, but it is large enough to take the small fox as well. The park staff has instituted a recovery plan that includes a captive island fox breeding program, the capture and relocation to the mainland of a significant portion of the golden eagle population, and the restoration of the native bald eagle to help displace the golden eagle.

CONSERVATION CONCERNS

Prior to 1800 the Channel Islands were the home of the Chumash Indians, living in scattered settlements. Their lifestyle was supported by the islands' shrubs, woodlands, and grasslands. Spanish-American settlements soon appeared, spreading from the mainland. The primary land use was ranching, some of which continued until 1998.

The legacy of the ranching era was severe and continues to cause problems for park management. Ranchers obviously prefer grasslands to other ecosystems; conversion of shrub and woodland to grassland was widespread. The presence of livestock caused significant soil compaction and subsequent soil loss. Weed species were introduced.

The impacts on wildlife were also severe. Many species were extirpated when their habitats were reduced or destroyed. Others, like the bald eagle, were extirpated more directly by harassment and shooting. The use of pesticides also contributed to the loss of native fauna.

Native wildlife is being harmed by introduced animals that ultimately became feral. These include pigs, sheep, horses, burros, and cattle. The infamous black rat also made its appearance. Ridding the islands of the progeny of these animals continues to be one of the park's major issues. There are some successes: burros have been removed from San Miguel as have sheep on Santa Cruz and pigs on Santa Rosa. The removal of feral pigs from Santa Rosa was the first time such efforts have been suc-

cessful over a large area. Conversely, feral pigs remain uncontrolled on Santa Cruz Island.

Some exotic wildlife was also introduced, including rabbits, elk, and deer. Elk and deer remain on Santa Rosa, being managed under special-use permit that is renewable through 2011. After that date, they will be removed.

Repairing the evidence of past human influences will likely be the park's major effort in the future. There are, however, other issues. The huge human population on the nearby mainland is a source of ozone and other pollutants. Santa Ana winds bring pollutants off shore from the Los Angeles basin. Offshore oil and gas extraction facilities are also a source of problems. Damage to vegetation and the loss of visibility have been noted on the islands. The potential for increases in pollution is related to population growth and a more widespread search for energy resources.

Regulations prohibit aircraft overflights of the islands below one thousand feet. Violations occur about once per week, but because of lack of staff, enforcement is difficult. Nesting of pinnipeds and seabirds is highly vulnerable to overflights. With pinnipeds, low overflights may cause stampedes that can crush young pups. They also cause abandonment of seabird nests. Also, the islands are just down range from Vandenburg Air Force Base from which the military launches its space vehicles. Launches cause frequent sonic booms over the islands.

Restoring the island's biodiversity will take much effort and significant funding. Two hundred years of human impact have caused huge changes in the island ecosystems, and a great deal of work remains to restore them to their previous glory.

CONGAREE
Primeval Forest of the Southeast

Hopkins, South Carolina
ACRES: 21,890

Many of America's national parks are literally known to everyone. Most Americans can identify Yellowstone and Yosemite National Parks, while other parks are lesser known. Perhaps the least known American national park is Congaree. Congaree National Park, located near South Carolina's capital city, Columbia, was established November 10, 2003. Congaree's claim to fame is the largest intact old-growth floodplain forest in North America. The area was originally designated Congaree Swamp National Monument in 1976 in order to protect the area from potential logging. National park status was achieved twenty-seven years later.

Some of the trees here are truly magnificent. Among the largest are a 162-foot cherrybark oak (24-foot circumference), a 145-foot loblolly pine (15-foot circumference), a 144-foot American elm (16-foot circumference), and a 124-foot water tupelo (21-foot circumference). In the low flooded areas, the primary trees are bald cypress and tupelo, while on stream banks you'll find sycamore. The most common forest in the park is of sweetgum and mixed hardwoods, including swamp chestnut, laurel oak, and green ash.

The history of the area revolves around the forest. During the American Revolution, Francis Marion, better known as the "Swamp Fox," used these forests to hide from the British. Later settlers in the area found agriculture very difficult due to flooding. Even logging didn't last. However, in 1969, higher timber prices allowed private landowners to reconsider logging. The Sierra Club responded with a successful grassroots program that led to federal purchase and designation as a national monument, and ultimately a national park.

There are two ways to see the park's 21,890 acres—on foot or by canoe. The visitor center provides access to boardwalks that give the casual visitor a way to see this incredible forest. For those wishing to see the more remote wilderness portions of the park, several trails exist, ranging in length up to 11.1 miles.

Canoeing is popular on the Congaree River and Cedar Creek. This allows the visitor an up-close look at some of the largest trees and an opportunity to see many of the park's mammals and other wildlife. You must bring your own canoe or rent one from outfitters in nearby Columbia. Some routes may be difficult at times due to flooding and downed trees.

Many of the mammals in the park are nocturnal and rarely seen. These include the bobcat, Virginia opossum, raccoon, southern flying squirrel, and the rare (in this area) fox squirrel. There are several others that are out during the day. The largest is the white-tailed deer. The only other ungulate in the park is the wild boar, an early import from Europe. The gray fox is also present and you may find one searching for its prey, which includes the eastern cottontail and the eastern gray squirrel. Congaree is also home to the river otter. Other mammals in the park include the muskrat, long-tailed weasel, mink, striped skunk, and eastern woodrat.

CONSERVATION CONCERNS

There is a significant portion of the Congaree River above the park. Especially during floods, it is possible that evidence of human activities washes into the park. One report describes three domestic pigs riding a log through the park during flooding. More insidious are the pollutants from farms, septic tanks, and industry. Park staff has little control over land uses outside the park and must rely on citizens to protect their public lands from harm.

Continuing growth in the region suggests the need for additional power generation, which means more burning of coal. The result can be a decline in air quality in terms of both pollutants and visibility. More automobiles mean more hydrocarbons in the air as well.

Acid rain and global warming caused by more distant polluting sources are also a concern. The southern Appalachians are receiving huge volumes of pollutants from places like the Ohio River valley and the Midwest. The entire southeast region faces a dilemma of how to meet the energy demands of the public while maintaining air and water quality.

CRATER LAKE
The Liquid Sky

Crater Lake, Oregon
ACRES: 183,224

Many people recall the eruption of Mount St. Helens in Washington. Many don't know that a similar event created beautiful Crater Lake in Oregon. About seventy-seven hundred years ago, the twelve thousand-foot Mount Mazama erupted, filling the sky with ash and other volcanic debris. So much material was expelled that the volcano collapsed back into itself, creating a volcanic basin known as a caldera. Crater Lake occupies Mount Mazama's caldera.

Crater Lake's caldera is characterized by sheer cliffs, some almost two thousand feet high. Because of the volcanic origin, there are no inlets or outflows of water at Crater Lake. Instead, the lake is fed by snow-melt from the huge snowfall (533 inches annually). Seepage and evaporation generally equal snowmelt. Lake depth varies by less than 1 percent per year.

Crater Lake is the nation's deepest lake, averaging 1,943 feet deep. Worldwide, it is the seventh deepest. The greatest width of Crater Lake is six miles; at its narrowest it is 4.5 miles across. Total lake area is 13,069 acres and its volume is five trillion gallons of water. Water quality is extremely high since no sediments flow into the lake, and most input is directly from precipitation.

The thirty-three-mile Rim Drive circles the lake providing beautiful vistas and access to camping and picnicking facilities. It is open only in summer. In winter, you can drive to Rim Village and walk to the rim through huge banks of snow.

Though the lake is 6,176 feet above sea level, some park peaks approach nine thousand feet. In addition to the lake, the park contains nearly 180,000 acres of volcanic topography including cinder cones, pumice flats, and lava. Forests are typical of the Cascade Range with various firs at higher elevations and ponderosa and lodgepole pines lower. Access to most of these areas is by foot only.

Mammals of Crater Lake are typical of the Cascades as well. The commonly seen animals are the rodents. Probably most common is the golden-mantled ground squirrel. In the forest, most common is Douglas's squirrel. Others are the western gray squirrel and porcupine. The California ground squirrel and the bushy-tailed woodrat are locally common as are the Townsend chipmunk, northern flying squirrel, and western (or Sierra) pocket gopher. The largest rodent, the beaver, has resided in the park in the past but is now rare, as is the muskrat.

At higher elevations, you may see the American pika. Other members of the hare family are the snowshoe hare and the mountain (or Nuttall's) cottontail.

Both mule and black-tailed deer (a subspecies of mule deer) are common to the park. The mule deer are most commonly found in the drier parts of the park; the black-tail is usually found in the wetter. The white-tailed deer lives just outside the park but is considered accidental within Crater Lake.

Another ungulate common in the park is the Roosevelt elk. It is resident in the park during summer but migrates outside the park during winter.

All this life at Crater Lake supports a wide variety of predators. Perhaps the most often seen is the red fox. The gray fox is rarely seen. The coyote is common, even in winter.

In most parks, it is very uncommon to see a long-tailed weasel, but at Crater Lake they are commonly encountered during daylight. A relative, the short-tailed weasel, is nocturnal and less common. The striped skunk is locally common, as are the marten and badger.

Mountain lions and bobcats may be resident in the park but are rarely encountered. Two other rare carnivores in the park are the wolverine and Canada lynx.

Finally, what of the black bear? The black bear is common in the park. The Park Service estimates that the park's bear population ranges between thirty and forty. Unfortunately, the population may be declining due to habitat loss and hunting outside the park. The concern of management is to maintain the bears' inherent fear of humans and to prevent them getting a taste for human food.

CONSERVATION CONCERNS

Air quality at Crater Lake is uncommonly high and has been compared to the quality of air in Antarctica. Visibility frequently exceeds one hundred miles. Even so, the future holds questions. Pollution from afar may bring acid rain and snow, causing a decline in the park's purity. Chemicals and particulates from nearby man-caused forest fires and other sources will

Averaging 1,943 feet deep, Oregon's Crater Lake is the nation's deepest and contains five trillion gallons of water. The lake was formed more than seventy-five hundred years ago when Mount Mazama erupted. Visitors to the lake's shore can see golden-mantled ground squirrels scurrying about at the edge of the woods.

bring nutrients into the lake, reducing the water's high quality. It is hoped that the park's monitoring system for air and water pollution will identify problems early enough that solutions can be found.

As in all parks, park management has little or no control over land use outside park boundaries. At Crater Lake, the decline in bear numbers is indicative of resource management issues around the park. We can only hope that decision makers on public lands around the park, as well as private landowners, will support the values of Crater Lake and the protection of its resources that is needed.

CUYAHOGA VALLEY

Clear Waters of the Crooked River

Brecksville, Ohio

ACRES: 32,861

June 11, 1969: the national media reports that the Cuyahoga River has caught fire. This wasn't the first time it happened. Chemicals and oil slicks flared regularly. In fact, the river had burned periodically since 1936. But this latest fire caused the nation to notice and contributed to public support for environmental legislation including the Clean Water Act and the creation of the Environmental Protection Agency under President Richard Nixon.

To restore the river, polluters had to be controlled and land use regulated. Several methods were used, one of which was for the government to acquire portions of the 813-square-mile watershed. No one agency would own the entire hundred-mile length of the river, but the federal government became very active. Perhaps the crown jewel of management of the Cuyahoga was created in 1974 when Congress established the Cuyahoga Valley National Recreation Area. The National Park Service acquired twenty-two miles of river frontage and adjacent lands totaling thirty-three thousand acres. The area extends from the southern suburbs of Cleveland south to Akron. The success in restoring the river resulted in the Cuyahoga being designated an American Heritage River in 1998 and the park redesignated as Cuyahoga Valley National Park in 2000.

Managing and restoring Cuyahoga Valley has not been easy. The area has a long history of human use. Today, it is an urban park crisscrossed by highways and railroads and coexisting with industry and municipalities. Though only 2 percent of the watershed is urban, 15 percent of Ohio's population lives here. The result is a fragmented ecosystem that includes forests, fields, and seventeen hundred widely scattered acres of wetlands. About seventy manmade lakes and ponds are found within the park. Even so, Cuyahoga Valley has become the home of considerable wildlife including thirty-two mammal species. Some species that had been extirpated are coming back.

The most widely seen mammal in the park is the white-tailed deer. This browser can be seen at dusk and dawn especially in wooded areas. The Park Service estimates that their population has been growing approximately 9 percent per year for the last twelve years. The population has grown to a point that a task force has been created to assess the impact the deer are having on surrounding communities and farms.

Once extirpated from the valley, the beaver has returned and is locally abundant in the park. They have caused considerable impact on the size and distribution of wetlands within the park. They are frequently seen in wetlands as is the muskrat.

Several small mammals are abundant in the park including voles, moles, and squirrels. The most frequently seen small mammal is the eastern chipmunk. Other mammals that are common to Cuyahoga Valley include the Virginia opossum, striped skunk, muskrat, woodchuck, and raccoon. Others present but uncommon are the southern flying squirrel, mink, gray squirrel, fox squirrel, red squirrel, and eastern cottontail.

In an urban environment such as this, you would not expect to find large predators. Indeed, none live here. Two species of foxes do live here, the common red fox and the less common gray fox. Recently, the coyote has made its way into Cuyahoga Valley, as it has in many parts of the Midwest. The least weasel and long-tailed weasel are resident but are rarely seen.

There are several bat residents in the park, some of them quite common. In July 2002, the federally endangered Indiana bat was found here as well. This was the first time it had ever been recorded in Cuyahoga Valley. The efforts to improve habitat for wildlife are apparently paying off handsomely.

CONSERVATION CONCERNS

The park's location in the midst of one of America's largest urban complexes causes serious concerns. The legacy of human abuse of the Cuyahoga River still exists. The highly developed areas upstream are the source of substantial pollution. During storms, huge volumes of runoff find their way into the river. Storm sewers are overwhelmed during heavy rain events.

In addition, not all water entering the river is completely disinfected, causing concern for human health. A recent study found that 98 percent of water samples exceeded federal standards for *E. coli*. Salmonella was found in 68 percent of the samples, and infectious enteroviruses in more than half. Consequently, the Park Service does not recommend the river be used for swimming, canoeing, or wading. Water quality goals for the river continue to be the remediation of current pollution

Containing sandy dunes and high mountain peaks, Death Valley is America's hottest place and at night also one of its coldest. Ground temperatures above 200°F during the day and below freezing at night have been recorded. The "elevation" of the park ranges from below sea level to over ten thousand feet above sea level.

sources, prevention of future sources, and the re-establishment of beneficial uses of the river such as swimming.

Air quality is also an issue here. Pollutants from Cleveland and Akron sometimes cause federal air quality standards to be exceeded. Ozone is a problem. Not only does it affect human health, it also causes problems for much of the vegetation in the park. Ozone alerts are sometimes issued, especially on hot summer days.

Relics of earlier land uses remain in the park. Mines, quarries, dumps, and industrial sites may contain unsightly and dangerous refuse. Direct action may be necessary to restore such sites. Lands formerly in agriculture are slowly reverting to more natural settings.

Disturbance also allows non-native species to become more easily established. Some of the more insidious of them already occur in the park including purple loosestrife, autumn olive, garlic mustard, and multiflora rose. A volunteer monitoring and control program is being developed, but ridding the park of these will be extremely difficult.

There are success stories at Cuyahoga Valley in restoring the river and its adjacent lands to create high quality wildlife habitat and a natural environment that makes outdoor recreation attractive. There is much more to do. In this urban setting, the problems don't get any easier.

DEATH VALLEY
The Land of Extremes

Death Valley, California

ACRES: 3,372,402

When you mention Death Valley, there is one word that comes to most minds—heat, legendary heat! The numbers are astounding. The maximum daily temperature at Furnace Creek exceeds 100°F from late May through September. In 1974, the temperature stayed above 100°F for 134 days. On July 15, 1972, the ground temperature at Furnace Creek reached 201°F; the air temperature that day was 128°F. Conversely, winter nighttime temperatures frequently drop to 15°F.

Annual average rainfall is only 1.70 inches. For one sixty-five-year period (1911–75), only 109.35 inches in total fell, less than one year's precipitation elsewhere in the United States. In 1929, absolutely no rain fell. The largest annual rainfall during that period was 4.54 inches, in 1913.

Not only does Death Valley have extremes in climate, there are also extremes in topography. The nation's lowest elevation is at Badwater, 282 feet below sea level. The highest point in the park is Telescope Peak (11,049 feet), only fifteen miles from Badwater. Even higher Sierra peaks are visible to the west of the park. This difference in elevation creates substantial differences in precipitation and temperature and consequently a variety of wildlife habitats.

Even so, people have lived in Death Valley for hundreds of years. Evidence of Native American campsites and trails remains in the park as do examples of primitive rock art. The first Anglo-Americans entered Death Valley on Christmas Day, 1849, seeking a shortcut to the California gold fields. Surprisingly, only one gold seeker died in Death Valley during the Gold Rush. Many of the unique place names in Death Valley come from these travelers.

Subsequent visitors were primarily prospectors seeking silver and other precious metals. The most noted mines were for borax, removed from Death Valley by the famous twenty-mule teams. Even though Death Valley was declared a national monument in 1933, borax mining continued until the late twentieth century. Finally, in October 1994, Death Valley was designated a national park. Ninety-five percent has been designated as wilderness.

With the variety of habitat comes a diversity of mammals: fifty-one species of native mammals live in Death Valley, most commonly rodents. The most common is the white-tailed antelope squirrel. The round-tailed ground squirrel is a common roadside resident. The Panamint chipmunk is resident in the pinyon-juniper forests. There are also two species of woodrat, the bushy-tailed and the desert, but both are nocturnal and rarely seen.

Two species of cottontail are resident. The desert cottontail lives in mesquite thickets in the desert, while the mountain cottontail lives at higher elevations. The nocturnal black-tailed jackrabbit's range overlaps with the cottontails.

Other mountain dwellers are the porcupine, western spotted skunk, and ringtail. The badger is also found at higher elevations but in the low desert as well. None of these species is often observed.

The coyote is common throughout the park and is best seen where there is a large rodent population. The nocturnal kit fox is also common throughout Death Valley. A third canid, the gray fox, is resident; its range is limited to the Panamint Mountains.

The other large predators in the park are the mountain lion and bobcat; neither is often observed. The mountain lion resides mainly in the mountains, while the bobcat is found from the desert to the surrounding mountains.

There are two large native grazers in the park. The mule deer is found in the pinyon-juniper forests in surrounding mountains but seldom in the desert. It prefers a habitat that is cooler than at lower elevations and better watered.

Most widespread in the park are the desert bighorn sheep. They live throughout the park at all elevations but remain in inaccessible cliffs and canyons. They are totally dependent on about forty seeps and springs where they have no competition with humans or domestic animals.

CONSERVATION CONCERNS

You are also likely to see feral burros and horses in the park. Both are the legacy of earlier human activities, mostly mining. Feral burros have been in Death Valley since the 1880s. These feral species cause serious impacts to native wildlife, especially the bighorn sheep. They consume forage the natives need and frequently foul waterholes. Bighorn sheep usually avoid any contact with burros and horses, causing the area they typically use to be severely restricted.

The Park Service has a plan to completely remove both burros and horses. A five-year capture program will be followed by two more years to let animal protection groups remove additional feral species. The final phase is to remove all the remainder using the most humane and cost-effective means available.

There are other non-natives in the park as well. Several exotic aquatic species live in some of the park's bodies of water. These include goldfish, crayfish, and bullfrogs. All likely have a negative effect on the native fauna such as the endemic Devils Hole pupfish. The objective for these animals is also eradication.

Though remote, Death Valley still suffers from air quality degradation. A nearby source of pollution is the Searles Valley, a dry lake bed that produces several chemical products. The result is dusty air with reduced visibility. More local causes of dust include off-road vehicles, agriculture, mining, and wind-blown soil.

During summer, prevailing winds bring pollutants from farther away. These southwest winds bring ozone and other chemicals from cars, power plants, and industrial facilities in the Los Angeles basin. The result can be acid rain and reduced visibility.

Unfortunately, park management has little or no control over pollution sources outside park boundaries. Pollutants will continue to be monitored and inventoried to document air quality. The National Park Service also seeks to achieve the highest level of protection under the Clean Air Act, Class I, which does not allow any further degradation in air quality by restricting concentrations of pollutants at current levels.

Water issues at Death Valley are a matter of quantity rather than quality. Water is a precious commodity in the desert, and there are many competing uses. Management is concerned about maintaining and restoring natural water sources in the park. Some wildlife could become extinct without sufficient water. It is likely that conflict related to water will be fought in the courts and within federal and state legislative bodies.

There are several other issues to consider. Death Valley is surrounded by military installations and military overflights occur frequently. The noise is highly disturbing to wildlife and visitors. Park staff is working with Defense Department officials to minimize such impacts.

DENALI
Home to Mount McKinley

Denali Park, Alaska

ACRES: 6,075,030

On May 7, 1932, an expedition led by Mount McKinley National Park superintendent Harry Like and Minneapolis attorney Alfred Lindley reached the south summit of North America's highest peak (20,320 feet). They were the second group to climb Mount McKinley successfully and the first to climb both the north and south summits. During their ascent, they found a recording thermometer left by the Stuck-Karstens expedition in 1913 during their pioneer climb. It read the minimum, –95°F, but the actual temperature was probably somewhat lower—the coldest reading ever in North America.

Mount McKinley is indeed a mountain of superlatives. Not only is it the highest and coldest in North America, it may also be the highest above the surrounding landscape, some 18,000 feet. It may also be the most difficult to see. In a typical summer month, the summit may be visible for only a day or two. Its extreme height creates its own weather, with the summit usually socked in by clouds. It is more visible in winter, but tourists are few at that time of year.

Mount McKinley was first viewed by Europeans in 1794 when British navigator George Vancouver noted a huge mountain to the north while aboard his ship in Knik Arm of Upper Cook Inlet, some 140 miles distant. It wasn't until 1878 that Arthur Harper reported seeing a "great ice mountain" while traveling on the Tanana River near present-day Fairbanks. It was Harper's son, Walter, who was the first human to reach the summit twenty-five years later.

The mountain was given the name McKinley in 1898 in honor of the Republican nominee for president, William McKinley of Ohio. McKinley never visited Alaska and never saw his namesake. Whatever Native American name the mountain had was soon ignored. In 1917 Congress established Mount McKinley National Park. While the mountain retains its "American" name, the name of the park was changed to Denali National Park in 1980, which means "The High One" in the local Athabascan tongue.

Nearly a million people come to the park each year. The summit is still the dream of thousands from all corners of the globe. Unfortunately, many climbing attempts have ended in fatalities. Others come to the park to see Alaska's splendor and to seek the opportunity to see the state's most charismatic wildlife. The park is on the major highway and railroad connection between Anchorage and Fairbanks, making Denali one of the most accessible parks in Alaska.

Mount McKinley is the chief feature of the park's landscape. The surrounding lowlands are covered with taiga, a Russian word for forests of primarily white and black spruce as well as birch, aspen, and larch. Higher elevations are characterized by tundra, an association of mosses and lichens with small shrubs. The park's six million acres have a diversity of habitats that support thirty-nine species of mammals.

The top predator is the grizzly bear. Most of the park's roads are closed to vehicle traffic except tour bus traffic where bears are frequently encountered. Some portions of the park's backcountry are closed to all visitor access to protect hikers but also to minimize impact on bears. The black bear is also resident in Denali and typically found in forested areas. As in most places where grizzlies are common, black bears are less often observed.

Among the ungulates, the moose is common everywhere in the park, including along the tour bus route. They are browsers so are found in forests and adjacent lands especially where willows are found. The caribou migrates through the park during summer. They winter north of the park but in May and June move through the park onward to the Alaska Range where they feed for two or three weeks. Dall's

sheep, a totally white close relative of the bighorn sheep, lives at higher elevations.

The gray wolf is relatively abundant but seldom spotted by visitors. The red fox is found throughout the park in all habitats. Coyotes are resident but seldom seen. Several members of the weasel family reside in Denali: the wolverine, river otter, American marten, ermine, least weasel, and American mink.

The snowshoe hare is commonly seen at Denali. Dependent on the snowshoe hare is the lynx, which exists only where adequate hares are present. They are few in number in the park so are seldom encountered.

The most common of the smaller mammals is the Arctic ground squirrel. In the spruce forests, you might see the red squirrel or porcupine. Around the park's lakes you may see the beaver swimming during the long summer days. Muskrats may be present as well. At higher elevations, the hoary marmot can be found. Another dweller of this habitat is the collared pika whose presence can be identified by its whistling voice.

CONSERVATION CONCERNS

The park's six million acres are mostly undeveloped, but will they always remain so? Concerns about human advances resulted in the National Parks Conservation Association's declaring Denali as one of America's ten most endangered parks in 2003. Their concerns included demands for unlimited snowmobile and ATV access to Denali, a proposal to build a ninety-mile road or rail line to Kantishna, a forty-acre commercial resort near Wonder Lake within the park, and a general lack of funding for Alaska parks. They noted that Yellowstone and Yosemite both have more employees than the entire Alaska Region.

For 2004, Denali was removed from the list for two reasons. No recommendations will be made this year about the proposed new road building, and Alaska's congressional delegation backed off on motorized access for now. Unfortunately, both issues could quickly reappear.

Denali has some of the cleanest air measured in the United States. Even so, contaminants are being found from Asia and Europe that come over the North Pole. Such pollution will likely continue to increase. The only response to it is through diplomacy and the treaty-making power of the federal government. Denali continues to be protected at the highest level under the Clean Air Act, Class I.

Water quality in the park is also very high. Airborne pollutants continue to be deposited but quality has not been affected.

Another concern is noise, one major source of which is scenic overflights. Flightseeing has increased four-fold in the past few years. It is the chief complaint of backpackers. Growth in the market is expected to put Denali among the most overflown national parks. There are legal limits to overflights, but Congress exempted Alaska from them, leaving park staff with little control. Aircraft and snowmobiles are combining to negatively influence the naturalness of the park.

Finally, unique to Alaska is the issue of subsistence. Subsistence is the taking of wildlife by rural residents in the portion of the park added in 1980. How these activities will affect the park's mammals remains to be seen.

DRY TORTUGAS
Nature of the Florida Keys

Key West, Florida
ACRES: 64,701

Most people think that Key West is the end of the keys with only ocean beyond. Not true. Accessible by ferry or air, Dry Tortugas National Park lies seventy miles west of Key West. Discovered by Ponce de León in 1513, Dry Tortugas consists of seven small coral reef and sand islands, surrounded by shoals and open waters. The name comes from the Spanish for turtle, *tortuga*, because of the large numbers of sea turtles present. The epithet "dry" comes from the fact that no fresh water exists on the islands. The park contains 64,701 acres, but only 104 of that is land.

Dry Tortugas has both natural and cultural resources to enjoy. The park contains one of the best developed tropical coral reefs in the Gulf of Mexico. The reef is inhabited by a wide range of coral species, many which are rare elsewhere. The reef also supports one of the best fisheries in the region.

The park provides critical habitat for sea birds. Many come to the islands to nest and support their young. In spring, thousands of sooty terns arrive to breed and nest. When the young are capable of flight, the terns disperse.

Dry Tortugas is also a significant nesting site for

sea turtles. The most common of them is the logger-head turtle. Also, thousands of green turtle hatchlings have been released here to increase their numbers.

There are significant cultural resources at Dry Tortugas. The largest Civil War–era coastal fort is here. Fort Jefferson was started in 1846 on Garden Key but never finished. The fort was garrisoned by the Union during the Civil War but saw no action. After the war, Fort Jefferson became a prison. The most infamous of its residents was Dr. Samuel A. Mudd, a member of the "Lincoln Conspirators." He was convicted of abetting John Wilkes Booth in his murder of President Lincoln by setting Booth's broken leg during his attempted escape. Three other conspirators were also imprisoned here. On the largest island, Loggerhead Key, a lighthouse was constructed in 1857 that is still operable.

Later, the islands became a naval base. It was a coaling station during the Spanish-American War and was the last port for the battleship USS *Maine* before the ship arrived in Havana for its rendezvous with destiny. Eventually the base became obsolete, and in 1935 Fort Jefferson National Monument was created. In 1992, the monument was greatly expanded and declared Dry Tortugas National Park.

In south Florida, nineteen species of mammals are commonly found. At Dry Tortugas, only two have been found based on sightings of live, non-stranded animals. These are the bottlenose dolphin and the manatee. The gentle "sea cow" eats marine vegetation in the shallow waters around the islands.

CONSERVATION CONCERNS

The isolation of the Dry Tortugas has helped protect it. There is no air pollution from industry or automobiles found here. Surprisingly, it has been thought that some dust is falling on the islands from the Sahara Desert. Similarly, there is little water pollution since there are dynamic ocean currents here and little land for runoff or sewage to be created. Ambient sound levels are extremely quiet and little light mars the night sky.

Conversely, visitation at Dry Tortugas continues to grow. Just how many people must be present before problems arise is not known at this moment. Continued monitoring of the natural environment is critical.

Finally, in a historical sense, Dry Tortugas National Park has been the site of numerous shipwrecks. These relics of history are important archeological resources. However, future shipwrecks may be different. Oil slicks and chemical spills are possibilities and could have profound impacts on the islands' natural environments.

EVERGLADES
The River of Grass

Homestead, Florida
ACRES: 1,508,537

Each day Florida's population increases by nine hundred people. Those nine hundred people need a place to live, a place to work, and food and water. That's nine hundred people 365 times a year. In addition, thirty-nine million people visit each year. What happens to them, or perhaps we should ask what happens to the wild places that may be needed to support these new residents and visitors?

Everglades National Park is a prime example. Located just to the west of the Miami metropolitan area, the park's 1.4 million acres of marshland would long since have been made into homes, farms, businesses, and industry without park protection in 1947. Even so, many believe Everglades is the nation's most at-risk park.

The Everglades are located at the southern end of a wide and shallow "river of grass" that begins in Lake Okeechobee. This "river" is as wide as fifty miles in places but is usually less than a foot deep. The water flows slowly through sawgrass over a substrate of limestone and ultimately empties into the sea. The maximum elevation is only eight feet in the park, making the gradient of flow almost flat.

The brick casements of Fort Jefferson heard the footsteps of Civil War prisoners including Samuel Mudd, convicted of setting the broken leg of Lincoln assassin John Wilkes Booth. Dry Tortugas is a mixture of historically significant buildings and incredible natural wonders, such as the enormous seabird colonies that live there.

The pineland ecosystem is an early successional stage in the Everglades. Park managers use prescribed fire to maintain these forests. The park is home to a variety of plants and animals, including the elusive Florida panther.

Wherever the elevation rises above the surrounding marsh, small islands, or hammocks, appear that may be covered with pine or tropical hardwoods.

The park also contains a portion of Florida Bay located between the upper Florida Keys to the east and the southernmost tip of Florida's mainland to the north. The saltwater bay is very shallow and characterized by the presence of numerous small islands or keys, some with water-accessible campgrounds. The most notable animal here is the American crocodile.

Everglades National Park is a wildlife paradise. Birdlife was one of the reasons for original protection due to the loss of species to plume hunters. Nonetheless, more than 90 percent of the wading birds have been lost. Reptiles are important as well. For many, the sight of an alligator is the highlight of their visit. Forty species of mammals are found in the park.

There are several furbearers common in the park. The marsh rabbit is the most abundant mammal. The round-tailed muskrat is locally common as are raccoons. The river otter is sometimes found in freshwater marshes. Opossums are commonly seen in hardwood hammocks and pinelands.

The white-tailed deer is also common throughout the park. They can be seen in the pinelands or running through the sawgrass of the freshwater marshes.

The most common predator is the bobcat, found in the pinelands and the hammocks. You may also see a gray fox in the pineland. The striped skunk is sometimes seen at Long Pine Key.

Squirrels, so common in other parks, are rarely seen in the Everglades. Residents include the fox squirrel and the eastern gray squirrel.

In Florida Bay, two marine mammals are often seen. By far the most common is the Atlantic bottlenose dolphin. Occasionally, the small pilot whale can be seen in Florida Bay and on the west coast.

Four of the park's mammals are on the Endangered Species List. Two of them are small and seldom seen: the Key Largo woodrat (a subspecies of the eastern woodrat) and the Key Largo cotton mouse.

The West Indian manatee lives in the salty inlets north of Flamingo and sometimes in Florida Bay. They spend winter months grazing in the shallow salty waters where they can be seen by visitors who maintain silence. Manatees are shy with people and prefer not to be disturbed. The biggest factor in the manatee's endangered status is power boating. Many manatees show the scars of collisions with boats and/or propellers, while others don't survive an encounter with a high-speed propeller. Interpretation and education are needed to ensure the survival of this gentle creature.

The other endangered mammal of great concern is a subspecies of the mountain lion, the Florida panther. The panther is the last known population of mountain lion east of the Mississippi River. It is perhaps the rarest and most elusive mammal in the park; estimates of its population range from thirty to fifty with only a portion of those actually in the park. Habitat loss is probably the biggest factor in their decline, but collisions with cars and mercury in their prey also contribute.

CONSERVATION CONCERNS

Early residents of south Florida viewed the Everglades as a worthless swamp that would have to be reclaimed to be of any value. Dredging began in the late nineteenth century and continued through the twentieth. In 1948 Congress authorized the Central and South Florida Project, which called on the Corps of Engineers to construct a series of levees, water-control structures, roads, and canals to manage water throughout the region. The Tamiami Trail (US Highway 41) serves as a dike; the only water that enters the park does so through the control structures along the highway.

Water is diverted for use by households, industry, and agriculture before it enters the Everglades, frequently leaving the park with insufficient water. When too much water falls, the excess is flushed into the park. The result for the Everglades is having too much or too little water, or not having the correct quantity in the proper season. For ecosystems that have evolved based on the rhythms of nature, upsetting the water cycle can be deadly.

Another problem caused by the removal of water for human uses is the reduction of fresh water in the limestone aquifers in the region. Pumping out the fresh water allows salt water to replace it. The result is ruining the remaining water as a water source as well as destroying plant life. Quality drinking water continues to be scarce; some people have resorted to drinking desalinated water.

Not only does agriculture take water, it also returns significant chemicals that end up in the park. Fertilizers (phosphates and nitrates) and pesticides find their way into park ecosystems. One especially worrisome contaminant is mercury, first found in park fish in 1989. Now it is found in many mammals and reptiles, including alligators and raccoons. One Florida panther was found dead with mercury concentrations that would be toxic to humans. The source of the contamination is unknown.

The introduction of exotic fauna and flora is a continuing problem going back to the nineteenth century. Plants have been brought to the park for use as ornamentals, food sources, or biological control. To date, 140 species of plants are considered exotic. The most serious of them are Brazilian pepper, melaleuca, and casuarina. Controlling these species has proven to be very difficult.

Exotic and feral wildlife are also problems. Exotic fish have become so established that they will likely never be eradicated. Former avian and reptilian pets are common in the park. Feral dogs and pigs seem to be increasing in numbers. Dealing with all these species is an ongoing battle with no assurance that victory will be attained.

All these issues were debated throughout the 1980s, and state and federal programs were enacted to repair the years of damage Everglades National Park has suffered. In 1988, the federal government sued the state of Florida and the South Florida Water Management District for failing to protect the quality of water entering the park from agricultural areas. In 1994, Florida passed the Everglades Forever Act, making it responsible for water quality cleanup. The Corps of Engineers and the South Florida Water Management District have been given the task of directing the Everglades Restoration Plan.

According to the National Parks Conservation Association (NPCA), the plan has faults: "problems range from project delays to weak rules." The ongoing concern for the park has it remaining on the NPCA's "Ten Most Endangered Parks" list. This political battle will continue; the outcome is uncertain.

GATES OF THE ARCTIC
Untouched Beauty

Bettles, Alaska
ACRES: 8,472,506

Robert Marshall was a leader in wilderness preservation. Marshall's U.S. Forest Service career culminated as the agency's recreation chief. As author of the early regulations that defined wilderness and wild areas, Marshall contributed mightily to the evolution of the idea of wilderness that ultimately led to the passage of the Wilderness Act in 1964.

Marshall's early career took him to Alaska, where he worked for ten years describing the plant communities of the Brooks Range. He became familiar with two peaks that straddle the North Fork of the Koyukuk River, Frigid Crags and Boreal Mountain, which he referred to as "the gates to the mountains." In 1980 when Congress created a new national park in Alaska's Arctic north, the name Gates of the Arctic National Park was selected. The park's 8.5 million acres (more than twice the size of Connecticut) are almost totally undeveloped. No roads enter the park, nor are there any developed campgrounds or other structures.

South of the Arctic Divide is found taiga, Russian for "land of little sticks." The small tree species include white spruce, aspen, and birch. Species at higher elevations include birch, alder, and willow. North of the Divide is a tundra ecosystem that runs all the way to the Arctic Ocean. The highest peak in the park is 8,510-foot Mount Igikpak. Numerous rivers are born in the park including six that have been designated as National Wild Rivers.

Gates of the Arctic is a vast extraordinary wilderness. Most visitors arrive by bush plane to camp, hike, or float one of the park's rivers. The Dalton Highway passes close to the eastern edge of the park, and some visitors may access the park by walking from the road.

Grizzly bears are found throughout the park. Here in the Brooks Range they are primarily vegetarians but feed on small mammals as well. They can often be seen excavating ground squirrel burrows. In season they will also take young moose and caribou. Most grizzlies are found in the alpine and tundra areas of the park. Conversely, black bears live in the forested areas in the southern part of the park.

Wolves are wide ranging in the park. They live in packs of usually six or seven wolves and travel over huge territories. They prey primarily on caribou. Wolf populations are unknown but estimated to be between 150 and 200. Wolf hunting and trapping are allowed by subsistence permit holders, but the size of the take is unknown. It is thought that wolf populations will, however, remain healthy.

The moose lives in the forested parts of the park, primarily to the south. However, forested portions to the north may also contain moose. They are frequently the prey of subsistence hunters from nearby villages, especially Anaktuvuk Pass.

The peaks of Gates of the Arctic are windswept and rugged, perfect habitat for Dall's sheep. These creatures prefer open tundra on steep slopes where they can watch for predators. It is estimated that there are between twelve and fourteen thousand sheep in the park, and are likely the most commonly seen large mammal.

Caribou range throughout the entire region and spend a great deal of the year in migration. They spend the winter scattered around the south slopes of the Brooks Range. In spring they begin a migration through the park, up the river drainages and across the divide to the North Slope. Here calving occurs in May and June. In August, they begin a southward movement that ends with the arrival in their wintering grounds. The size of the migrating herd can be astounding. Some may contain thousands of caribou. Like the moose, the caribou is frequently taken by subsistence hunters.

Extirpated from the park area in the mid-nineteenth century, the musk ox was reintroduced into northeastern Alaska in 1969 and 1970. It rapidly expanded its range and soon was resident in Gates of the Arctic. Today, they are most often seen in alpine areas where wind keeps snow levels low.

Several furbearers are resident. In the park's forested areas to the south, the lynx and marten are found. The red fox and Arctic fox are residents. There are several members of the weasel resident in Gates of the Arctic including the mink, wolverine, and least weasel. Aquatic mammals are limited due to lack of suitable habitat. These include the river otter, beaver, and muskrat. Marten, lynx, beaver, fox, and wolf are frequently taken by subsistence trapping.

The lower part of the food chain includes thousands of smaller mammals. Most are shrews and voles but there are also Arctic ground squirrels, red squirrels, Alaska marmots, and Alaskan hares. The large predators need a huge quantity of these small prey animals for survival.

CONSERVATION CONCERNS

The remoteness of Gates of the Arctic has prevented much human impact. Fewer than a thousand visitors arrive each year, so direct impacts are few. Air quality and water quality are high since little development is nearby. Lack of access keeps motor vehicles out of the park.

Even so, the future is less clear. The oil industry is well entrenched on the North Slope and its potential expansion into the Arctic National Wildlife Refuge will bring many changes. The nation's quest for energy resources will continue unabated. Where oil and gas are to be found in Alaska is unknown. Getting energy to market is also an issue, especially in Alaska where pipelines have already caused considerable environmental impact. More pipelines mean more impact, but no one knows if Gates of the Arctic will be affected.

GLACIER
A long-standing tradition

West Glacier, Montana
ACRES: 1,013,572

The celebration of the bicentennial of the Lewis and Clark expedition helps us recall their extraordinary journey. We read their journals and see the west through their eyes written at a time when few Europeans had ever seen the lands acquired through the Louisiana Purchase. On their return journey, the two explorers split up for a time. Lewis led a party to explore the headwaters of the Marias River. They soon returned due to fear of the Blackfeet who inhabited the region. Before turning back Lewis

Gates of the Arctic is one of America's least visited parks. Only a fortunate few visit the peaks of the Brooks Range and see what forester Robert Marshall called "the gates to the mountains." This vast park, measuring more than twice the size of the state of Connecticut, was created by the U.S. Congress in 1980.

wrote, "The course of the mountains still continued from southeast to northwest; in which last direction from us, the front range appears to terminate abruptly at a distance of 3 miles." The first written words describing Glacier were recorded.

Those mountains that Meriwether Lewis saw are today some of the most splendid in the Rockies. Glacier's one million acres combined with Canada's Waterton Lakes National Park make up the Waterton-Glacier International Peace Park. The two parks are bisected by the Continental Divide. West of the Divide, forests are wetter due to weather systems from the Pacific, and include trees such as western red cedar, grand fir, western hemlock, and western larch. Forests on the eastern slope are drier and are more like forests in the southern Rockies; lodgepole pine and aspen are common. In alpine areas, the common trees are willow, birch, and fir.

Most of the park is accessible only by trail. There are more than seven hundred miles of trail, many of which are above timberline. The Going-to-the-Sun Highway bisects the park and crosses the Continental Divide at 6,646-foot Logan Pass.

Rodents are common including the Columbian ground squirrel. Much less common are the golden-mantled ground squirrel, the thirteen-lined ground squirrel, and Richardson's ground squirrel. There are two species of tree squirrel in the park, the red squirrel and the northern flying squirrel. Three species of chipmunks are common in Glacier, the yellow pine chipmunk, the red-tailed chipmunk, and the least chipmunk. The bushy-tailed woodrat is very common throughout the park especially in rocky areas and in old buildings.

Beaver are common throughout the park in lakes, ponds, and streams wherever there is sufficient food. Muskrats live mostly in ponds and streams on the west side.

A visit to the park's high elevations may bring an encounter with the American pika. Other lagomorphs include the snowshoe hare and the white-tailed jackrabbit.

The weasel family is common in Glacier. The largest of them is the wolverine. The marten is found throughout forested areas; the mink is common along streams and lakes. Three species of weasel reside in Glacier: the short-tailed weasel, the long-tailed weasel, and the least weasel. Badgers are common in the grasslands.

Glacier is home to both grizzly and black bears. Both live in spruce-fir forests, slide areas, and alpine meadows, but it is not uncommon for both to enter campgrounds. There are an estimated five hundred black bears in Glacier and two hundred grizzlies.

Glacier's predators include the coyote, which inhabits the park at all elevations below timberline. Several hundred exist in the park. The gray wolf originally lived in Glacier but by the 1940s had been extirpated. In the 1980s, Canadian wolves reentered the park and today several breeding packs inhabit the park. There are three felids, the mountain lion, bobcat, and lynx.

There are between fifteen hundred and two thousand mountain goats in Glacier. Goats live at high elevations, especially above timberline. Bighorn sheep are found above timberline in summer too, but move to lower elevations during winter. Between three and five hundred inhabit Glacier.

Four species in the deer family are found in Glacier. The most abundant is the mule deer. The white-tailed deer is also a resident. The largest ungulate in the park is the moose. There are a few hundred moose in Glacier, mostly along rivers and lakes on the western side of the park. Elk are scattered throughout the park in summer and less often seen.

CONSERVATION CONCERNS

Glacier lies far from large sources of population, so

air and water quality might be expected to be less problematic. However, the region containing Glacier continues to develop. The growth to the west of the park is substantial, characterized by expanding housing developments, resource extraction, and recreation in the private sector. Though the park itself is protected, encroachments harm the surrounding ecosystem, with fragmentation an especially difficult problem. Several wildlife species are at risk. The grizzly population is considered out of danger and may even be removed from the threatened species list. Nonetheless, monitoring must continue so that

Going-to-the-Sun Highway can be seen here as it climbs toward the summit of Logan Pass. One can stop at the visitors' center and see mountain goats milling about.

potential problems can be addressed quickly should they arise.

Wolves face similar problems. Distemper and parvovirus are in the wolf population, but the biggest cause of death is illegal killing by humans. The long-term survival of the wolf in the Glacier ecosystem will probably be based on the attitudes of the public in the region, especially those with a financial concern, such as ranchers.

The lynx is not currently listed as threatened or endangered, but listing was proposed in 1998. Their population numbers are variable, based primarily on the numbers of snowshoe hares available. Even so, since the 1960s the number of lynx sightings at Glacier has continued to decline.

Air quality continues to be monitored for contaminants and visibility. To date, air at Glacier is better than at most parks in the United States. When testing for ozone, results show that the park's resources are probably not affected. Regarding visibility, there is some impairment caused by particulate matter from forest fires and wind-blown dust. Again, continued monitoring is a must.

GLACIER BAY

A Marine Wilderness

Gustavus, Alaska

ACRES: 3,283,246

In 1794 Captain George Vancouver brought his vessel HMS *Discovery* into Icy Strait. He had been here before as a crew member on Captain James Cook's 1778 voyage on HMS *Resolution*. Cook saw and named Mount Fairweather, Glacier Bay's tallest peak (15,320 feet). Some would call Vancouver the discoverer of Glacier Bay, though native Tlingit people knew the area well long before Vancouver's visit. What Vancouver saw was only a five-mile indentation in the coast, an indentation that was surrounded by ice "as far as the eye could distinguish."

The man who told the world about Glacier Bay, however, was John Muir, the noted naturalist. Muir came to Glacier Bay in 1879, touring in a dugout canoe. By then, the glacial ice had receded forty miles. He recounts his first visit in his book *Travels in*

Alaska, published just before his death in 1914. One chapter, "The Discovery of Glacier Bay," describes the bay's glacial history. The writings of Muir and others made the public aware of Glacier Bay, ultimately leading to considerable tourism.

Though Glacier Bay contains spectacular mountain coastal scenery, ecology was the basis for its protection. The retreat of glaciers dramatically displays plant succession after the ice has receded. Scientists called for national monument status for Glacier Bay to protect this unique ecosystem; success was achieved in 1925. In 1980, Glacier Bay was upgraded to national park status under the Alaska National Interest Lands Conservation Act.

Glacier Bay is characterized by huge ice fields that are the source of glaciers that flow to the sea. These glaciers have advanced and retreated for thousands of years. Mountains like Fairweather tower above the ice. Glacial valleys were created and, where glaciers have retreated, forests of spruce ultimately appeared.

The sea is a major force here. No point in the park is more than thirty miles from salt water. Tides can change as much as twenty-five feet in six hours; tremendous waves sculpt the coast. Sea mammals are common and cause considerable excitement for visitors, but land mammals contribute to the visitor's experience as well.

The humpback whale migrates to Glacier Bay in June and stays through the summer. These baleen whales can be found in most fjords except where the waters are filled with glacial silt. Two other baleen whales are less common, the finback and the Minke whale. The gray whale resides on the outer coast during spring and fall migrations.

In near shore waters, killer whales (orcas) can be found in all seasons. The harbor porpoise is common throughout the park especially in the southern part of the bay. The Dall's porpoise is less common but often seen in Icy Strait.

There are several resident pinnipeds in Glacier Bay. The most common of them is the harbor seal. Unfortunately, its numbers have declined in recent years. Less numerous is the northern sea lion (also called Steller sea lion), but it is still widespread in the park.

Sea otters are thriving in Glacier Bay today. They were extirpated from the area in the nineteenth century but were reintroduced in the 1960s. Their population is growing, and they are frequently seen throughout the southern portion of the park.

As plant succession continues from south to north, habitats change as well. It is estimated that thirty species of land mammal exist in the park but their distribution is in continual flux.

Several members of the weasel family live in Glacier Bay. The marten is common in the forested parts of the park. Wolverines live all over the park but are rarely encountered. River otters can be found in both fresh and saltwater habitats. Ermine (also called short-tailed weasels) are common but rarely seen.

Forest dwellers include the red squirrel, northern flying squirrel, and porcupine. The hoary marmot resides in rocky areas at higher elevations. The beaver is a seaside dweller, found especially along Icy Strait.

There are three widely scattered ungulates in the park. At lower elevations, the moose occurs widely in open forests especially among willows. Moose are relatively new residents in the park, arriving only after suitable habitat was created through succession. Sitka black-tailed deer (a subspecies of mule deer) are found along Icy Strait. The mountain goat lives in high elevations where sufficient food exists and terrain allows for spotting predators.

Their wariness is warranted; several large predators can be found in the park. Both grizzly and black bears live in Glacier Bay. The grizzly is most common in the northern part of the park, while the black bear is found in the forested southern part of the park. Both may be seen anywhere in the park.

Wild canids are widespread at Glacier Bay. The range of the gray wolf seems to be spreading northward as vegetation changes. The coyote is also found throughout the park, though it is most likely found in the forested areas. Like the wolf, the coyote's range is increasing.

Another animal with an expanding range is the snowshoe hare. Lynx prey on them frequently, so the lynx range is growing as well. It is widespread in the park but rarely encountered by visitors.

CONSERVATION CONCERNS

Glacier Bay National Park is typical of Alaskan parks in that its remoteness protects it somewhat from air and water pollution. Contaminants are present but are not yet a major problem.

A major issue in the park is the cruise ship industry. Operating requirements and a limit on cruise ship numbers have been in place since 1979. A permit system was enacted in 1985, which allowed 107 cruise ship entries into the bay annually. In 1996, the quota was expanded to 139 entries (and perhaps higher) if monitoring suggests environmental impacts are controllable. The concerns about cruise ships have to do with the possibility of disturbing wildlife, especially humpback whales. Evidence suggests they are disturbed by noise. The potential for air and water pollution also exists. The outcome of

this confrontation is unknown.

Similarly, operators from nearby communities have attempted to establish flightseeing businesses over the park. Most have not been successful, but many continue to make the effort to reach profitability.

In 1976 Congress passed the Mining in the Parks Act, which essentially prohibits mining in most national parks. At Glacier Bay, most mining ended, with one exception. The Brady Glacier mineral deposit contains a significant quantity of copper-nickel sulphides. The deposit is actually under the glacier and is in private ownership. Whether it is ever developed will depend on social, economic, political, and strategic considerations.

GRAND CANYON
Beauty of the Colorado

Grand Canyon, Arizona
ACRES: 1,217,403

In 2003, nearly 4.5 million people visited Grand Canyon National Park, one of the most popular parks in the United States. The Grand Canyon, however, wasn't always so popular with travelers. The first Europeans to see the canyon were a group

"Where glaciers meet the sea" is given meaning in this photo. Tour boats allow visitors to approach the face of these glaciers. On rare occasions, icebergs "calve," breaking off huge chunks that drift into the water.

of Spaniards led by Garcia Lopez de Cardenas, sent by Coronado in August 1540 to find the Colorado River. They were successful finding the rim, but found the Colorado to be an extreme impediment to travel. Cardenas sent three agile young men to find a way to the river but they were only able to get a third of the way down. The written record doesn't allow us to identify exactly where they were on the South Rim. The party departed disappointed at not being able to go farther.

More than three hundred years later, the first Americans visited the canyon. Of course, Native Americans knew the canyon well, including sites along the river. In 1857 Lt. Joseph C. Ives led an expedition to explore the Colorado from below to determine how far a steamboat could go up the river. He entered what today is the lower part of Grand Canyon National Park. Apparently he wasn't greatly impressed. Ives described the canyon in his "Report upon the Colorado River of the West: Explored in 1857 and 1858":

The region is, of course, altogether valueless. It can be approached only from the south, and after entering it, there is nothing to do but leave. Ours has been the first, and will doubtless be the last party of whites to visit this profitless locality. It seems intended by nature that the Colorado River, along with the greater portion of its lonely and majestic way, shall be forever unvisited and undisturbed.

Ives' prognostication didn't come true. Not only do millions visit the rims, more than 115,000 people followed Ives on the river's surface, mostly by raft.

Grand Canyon National Park is characterized by an awesome chasm that has become world renowned. Looking down from the rims or hiking a trail into the canyon provides an extraordinary experience for visitors. Float trips and burro rides offer even more excitement.

The natural environment at Grand Canyon provides an interesting backdrop. The park's 1.2 million acres contain several ecosystems. Due to an elevation difference of eight thousand feet within the park, one can find ecosystems typical from Mexico to Canada. The South Rim is characterized by forest. At the lower elevations, the primary trees are juniper and pinyon pine with sagebrush and grasses interspersed. At higher elevations, ponderosa pine is most common.

The North Rim is about one thousand feet higher in elevation. Near the rim, ponderosa pine is common. At higher elevations behind the rim, spruce, Douglas-fir, white fir, and aspen are the primary forest cover.

Inside the canyon the desert predominates. The life zones are referred to as the Upper and Lower Sonoran zones, the latter of which is similar to what one would find in the Mexican state of Sonora. Plants and animals here are typical desert dwellers. The exception is along the Colorado River where the river's riparian zone supports more water-dependent plants and animals.

These life zones support varying habitats and, in turn, a variety of species. There are ninety-one species of mammals in the park. Some are found on only one rim; others are endemic to the park.

The most common animal group is the rodents. The most widespread is the cliff chipmunk, found on both rims and the upper portion of the inner canyon. There are several more rodents seen only on the North Rim: the Uinta chipmunk, least chipmunk, golden-mantled ground squirrel, and red squirrel.

Perhaps the most handsome rodent is the Abert's squirrel, also called the tassel-eared squirrel. Although once considered a separate species, the North Rim Kaibab squirrel is a subspecies of the South Rim Abert's squirrel. Both are dependent on ponderosa pine for survival.

The raccoon is common in canyon bottoms on both sides of the river. The ringtail is found along the Colorado but also up to the rim. The river otter may exist along the river but the cold water probably limits its presence.

The park's only common lagomorph is the desert cottontail on the South Rim. On the North Rim, the mountain (or Nuttall's) cottontail is resident. The uncommon black-tailed jackrabbit lives on both rims in a variety of habitats.

The weasel family is well represented at Grand Canyon. The western spotted skunk is found in the inner canyon and the striped skunk in the higher forests. The badger and long-tailed weasel are residents.

Perhaps the most widespread predator is the coyote. They can be found from river to rim but are more common higher. The gray fox is found from river to rim as well. Black bears are occasionally seen in the park though they may not be resident. They are usually seen in forested areas.

The bobcat is common throughout the inner canyon and on the rim. The mountain lion is also resident in a wide range of habitats but is considered rare. The attempt to remove them from the North Rim after World War I is one of ecology's horror stories.

The most common large mammal in the park is the mule deer. It is seen in most habitats including around the hotels in Grand Canyon Village. Uncommon in the inner canyon is the desert bighorn sheep. Its population has been affected by the presence of feral burros. About five hundred burros were removed in the 1980s, but concern still exists for the long-term survival of the desert bighorn. Elk occasionally enter the park from the south but are not resident. Finally, pronghorns are occasionally seen in the sagebrush and grasslands of the South Rim.

CONSERVATION CONCERNS

Issues at Grand Canyon National Park are a function of the huge crush of visitation as well as land use demands throughout the region. Annual visitation exceeds four million, causing impacts on soil, air, water, and wildlife. A new shuttle system has reduced the impact of the automobile by separating visitors from their cars, but crowding remains. Bikeways and walkways will ultimately be built, helping to reduce auto emissions.

Grand Canyon National Park has some of the cleanest air in the country. In addition, the park is protected by the highest protection possible under the Clean Air Act. Even so, population growth in neighboring states as well as Mexico suggests air pollution could get worse. These people will demand more and more energy, much of which will come from the burning of coal.

A problem already occurring in the park is the reduction of visibility. In a park where scenic beauty is such an important aspect of park values, visibility is crucial. Haze happens frequently in the park even though ambient air usually doesn't exceed standard. The average visibility is 106 miles but sometimes is reduced to fifty miles. Monitoring visibility will help us determine how to maintain air quality.

The major issue regarding water has to do with quantity. The completion of Glen Canyon Dam in 1966 caused significant changes in the Colorado River. The temperature of the river is much colder than before the dam, changing the species of fish and other animals that can survive in the river. As a result, some fish species have become endangered. Siltation has been reduced, making river water clearer, but the riparian ecosystem has changed as well. Large floods no longer occur, preventing the flushing action the river used to provide.

Another water problem is the availability of water for park uses. The water supply for the park comes from Roaring Springs on the North Rim; it is then piped to both rims. The pipeline to the South Rim may not have enough capacity to meet future needs. In addition, developers of motels, hotels, restaurants, and aircraft services in the village of Tusayan at the southern edge of the South Rim have asked for water from the park system in the past and will likely do so again in the future. National Park staff plan to set minimum flow requirements for Bright Angel Creek, the outflow of Roaring Springs. To do so will require that the withdrawal of water from Roaring Springs be curtailed and alternative sources found.

The presence of wild burros continues in the park, and their impact on desert bighorn has been documented. The bighorn has not been listed on the Endangered Species List yet, but to protect its habitat the burro must be removed.

Finally, an extremely popular way to see Grand Canyon is by aircraft—fixed wing or helicopter. Grand Canyon Airport in Tusayan has become one of the busiest airports in Arizona. Others embark in Las Vegas. For 1998, the estimate of use was 642,000 passengers in approximately ninety thousand flights. Even though there are restrictions on flights such as flying below the rim or flying over certain parts of the park, aircraft noise has become pervasive in most areas of the park. An aircraft noise management plan is due by 2006 with full implementation by 2008. Achieving such a plan will take considerable political skill.

GRAND TETON
The Stunning Mountains of Wyoming

Moose, Wyoming

ACRES: 309,995

Colorful characters have inhabited the American west. Some of them just couldn't seem to get enough adventure in their lives. A member of the Lewis and Clark expedition, John Colter, faced incredible hardships on the journey to Fort Clatsop, Oregon. As the party descended the Yellowstone River on the return portion of the expedition, they met a party of trappers going upstream. The lure of the mountains called to Colter, and he bade the prospects of civilization farewell. Most believe he spent the winter of 1807–08 in Jackson Hole, today part of Grand Teton National Park. We wonder what Colter thought of the majestic mountain range to the west. Trappers were followed by ranchers and in turn by today's tourists.

The Tetons are visually extraordinary. There are twelve peaks that exceed twelve thousand feet in elevation; the tallest is the Grand Teton (13,770 feet). They seem even higher, rising seven thousand

The peaks of the Tetons tower over the sagebrush of Jackson Hole. Trails provide access to the narrow canyons between the rocky spires. Twelve peaks of the Grand Tetons exceed twelve thousand feet in height.

feet above Jackson Hole. Treeline here is about ten thousand feet where Engelmann spruce, subalpine fir, and limber pine are found. Lower, the most common tree is the lodgepole pine. Along streams, aspen, cottonwood, and willow provide excellent cover for wildlife. Open meadows and sagebrush flats cover the valley. These habitats support exceptional opportunities for wildlife viewing including fifty-four species of mammals.

The most numerous mammal group is the rodents. The park's largest ground squirrel is the yellow-bellied marmot. Others include the golden-mantled ground squirrel and the Uinta ground squirrel. Three species of chipmunk are found throughout the park: the least, Uinta, and yellow pine chipmunks. The bushy-tailed woodrat is also present but uncommon.

The tree squirrels in the park include the abundant red squirrel and the northern flying squirrel. The porcupine is found in coniferous forests.

The park's waters are inhabited by rodents too. The beaver is the most common; muskrats live in most of Grand Teton's ponds.

Hikers in the high country frequently encounter the American pika. In the coniferous forests, the snowshoe hare resides. In the valley, the white-tailed jackrabbit can be seen.

Predators are often seen in Grand Teton National Park. There are several members of the weasel family common to the park: the marten, long-tailed and short-tailed weasel, badger, and river otter.

In 1998, two years after reintroduction into Yellowstone National Park, the gray wolf returned to Grand Teton National Park after a fifty-year absence. It is still considered uncommon in the park but is apparently expanding its range in the lower elevations in the valleys and river corridors. The coyote is abundant and widespread.

Grand Teton is the home of two species of North America's bears. The black bear is common especially in forested areas. The grizzly bear is more common on adjacent national forest lands, especially in the Teton Wilderness.

The lynx lives in wooded areas, while bobcats and mountain lions live in most park habitats. Chances of seeing any of these predators are unlikely.

Perhaps the highlight of most visitors' park experience is seeing one of the several large hoofed mammals. Several species can commonly be seen from park highways. The moose is abundant and is found in several habitats. Elk are abundant in the park as well.

Pronghorns, mule deer, and bison are also common in Grand Teton. Pronghorns are seen in the sagebrush areas; mule deer can be found in meadows and sagebrush flats. Bison are typically mammals of the prairie.

There are a few bighorn sheep in the park. They spend the summer in the high country. They migrate to lower elevations during winter and may sometimes be encountered in Jackson Hole.

CONSERVATION CONCERNS

The Teton Range was the attraction to create a very popular national park. Nearly 2.5 million visitors entered the park in 2003. But the area around the park exhibits substantial growth as well. Jackson, Wyoming, has been a major tourist attraction for several years. Now people see Jackson as a great place to live. What will the resulting crowds be like? Will traffic jams be part of the park experience?

With tourism soaring, transportation issues will become important. The Jackson airport is actually within the park. Several years ago a proposal was made to extend the runway in order to accommodate larger jet aircraft. The proposal was rejected at that time, but future increases in regional population may cause the issue to be resurrected.

Grand Teton also faces the typical impacts of western parks. Population growth around the park may cause both air and water pollution. National parks are designated for forever. What will Jackson Hole look like in a hundred years? Will mountain views still predominate or will suburbia be the rule?

GREAT BASIN
Sagebrush and Summits

Baker, Nevada

ACRES: 77,180

Many travelers do not understand the difference between a national park and a national forest. National parks are managed for protection and use of scenic and historic areas, while national forests are managed for multiple uses including timber production, grazing, wildlife, water, and recreation. The confusion increases when one is created out of the other. Such was the case with Great Basin National Park in Nevada.

The Great Basin is a two hundred thousand-square-mile area that contains all of Nevada and parts of surrounding states. This huge area is off the beaten path and was the last region in the west without a national park. Proposals to create a national park began in the 1920s. A small part of the area was protected in 1922 when President Warren Harding declared Lehman Caves National Monument, a 640-acre parcel on the flank of Wheeler Peak, to be managed by the U.S. Forest Service. In 1933, management was transferred to the National Park Service.

Efforts to create a Great Basin National Park continued. Finally, in 1986, Great Basin National Park was established, using 13,063-foot Wheeler Peak as the focal point. The 77,180-acre park was created by combining Lehman Caves National Monument with portions of the Humboldt National Forest.

The park includes the rocky summit of Wheeler Peak and the surrounding sagebrush flats. Just below treeline, limber pine, Engelmann spruce, and Douglas fir are common. At the same elevations, south of the summit of Wheeler Peak, one can find bristlecone pines, the earth's oldest living things. Great Basin's bristlecone pines exceed three thousand years of age. Forests of aspen, pinyon pine, and juniper cover the lower slopes.

A road climbs Wheeler Peak up to the ten thousand-foot level. Trails radiate from the road, including one to the summit. Access and the multiple habitats provide significant opportunity to see many of the park's sixty-one mammal species.

Great Basin has an abundance of rodents. In the montane forest habitats, the Uinta chipmunk and golden-mantled ground squirrel are frequently encountered. In the pinyon-juniper forest, the most common rodents are the tree-climbing rock squirrel and the cliff chipmunk. The least chipmunk, yellow-bellied marmot, and bushy-tailed woodrat are also here. At the lowest elevations, there are several common rodents including the white-tailed antelope ground squirrel, Townsend's ground squirrel, northern pocket gopher, and desert woodrat.

Among the lagomorphs, the black-tailed jackrabbit, mountain (or Nuttall's) cottontail, and desert cottontail are residents.

Predators follow the rodents. Both the western spotted and striped skunks are widespread in forested habitats. The long-tailed weasel and the short-tailed weasel are common in the woodlands. Great Basin has some of the larger predators too. The coyote is common everywhere in the park. The park's foxes are the gray fox and the kit fox. Both bobcats and mountain lions are residents.

Two hoofed mammals are common in the park, and two more are making a comeback. The mule deer is abundant in the mountains and foothills. The pronghorn is common in the sagebrush areas of the park.

The elk and bighorn sheep were original residents of the Wheeler Peak area, but both were extirpated in the twentieth century. Elk were reintroduced to the region in the 1930s. They are now in the park using woodland habitats. Several elk have been fitted with radio transmitters to assess numbers and habitats used. One recent count was seventy-

Bristlecone pines are among the world's most ancient residents, some being thousands of years old, and many show large areas of "dieback." These pines are found in the higher elevations of Great Basin's Wheeler Peak.

four individuals—a significant increase since 1999. The bighorn was reintroduced in 1979–80 but its population has showed little growth; they are rare in the southern Snake Range.

CONSERVATION CONCERNS

As a former national forest, Great Basin faced some unique problems caused by Forest Service management. National forest management allows grazing, logging, and mining, and evidence of these earlier uses still exists. In fact, when the park was created in 1986, grazing was allowed to remain in perpetuity as part of the park's enabling legislation. Visitors, however, didn't approve of livestock as a part of their experience. After years of effort, the Conservation Foundation, aided by Senator Harry Reid of Nevada, amassed enough funding to compensate ranchers who were willing to donate their grazing permits to the park. Once the park possessed the donated permits, they could be terminated. Sheep grazing, however, still continues in parts of the park.

Great Basin is a remote park and therefore has fortunately been spared significant air pollution. Visibility measurements in the park have been among the cleanest readings in the entire nation. Whenever winds come from the northeast, visibility suffers. The source of the pollution is thought to be Salt Lake City and a power plant in Delta, Utah.

Ozone and acid deposition do not seem to be a problem at present. Ozone levels are within standards, and acidity of the park's snow and rain is normal. Problems are not apparent now but certainly could become so quickly. Continued monitoring programs can help identify problems before they become severe.

Great Basin suffers from a malady common to many parks: exotics. There are more than twenty-five species of non-native plants in the park. There is not enough funding to treat all of them, nor is it necessarily possible to totally eradicate exotic plants. Visitors frequently are the means of spread for exotics, and interpretive programs should be used to make visitors aware of ways to prevent spread.

So far, the Great Basin region remains lightly populated, and residents cause little impact. The park receives very little use, with only eighty-five thousand visits in 2003. Yet the park's popularity is only likely to increase. More people mean more impacts, unless management remains vigilant.

GREAT SAND DUNES

Sand in the Shadow of Snow

Mosca, Colorado
ACRES: 84,670

Since the creation of Yellowstone National Park, the national park system has continued to grow. Nationally significant natural and cultural resources may be considered for inclusion, and with the approval of Congress and the president, officially become a unit of the National Park Service. The newest unit of the park system was officially dedicated in September 2004—Great Sand Dunes National Park.

Great Sand Dunes was originally designated as a national monument in 1932. In November 2000, the Great Sand Dunes National Park and Preserve Act was passed, authorizing the creation of the new national park once the Secretary of the Interior determined that sufficient land additions had been acquired. Some of the new park lands would come from a transfer from the Rio Grande National Forest. Additional lands would be acquired through the purchase of the privately owned Baca Ranch. The purchase of the ranch was completed on September 10; the Secretary signed the designation of the new park three days later.

The original reason for creating Great Sand Dunes was the presence of North America's tallest sand dunes. The park contains thirty-nine square miles of inland sand dunes. One would expect to find dunes along a sea or lake shore. But why here in the middle of the continent? Where did the sand come from? Deposits from mountain streams and playa lakes in the San Luis Valley provide the source. The prevailing southwesterly winds carry the sand grains toward the Sangre de Cristo Mountains. Upon reaching the mountains, reverse winds from the peaks overcome the sand's migration, causing it to accumulate at the base of the range.

The highest points in the dunes are Star Dune (rising 750 feet above the surrounding landscape) and High Dune (700 feet above). There are no established trails on the dunes, but many visitors enjoy a hike, sometimes to the top. Visitation for 2003 was 249,923.

The new park contains a wide diversity of life zones. One can walk from the dune field to the alpine zone within a day's walk. In the preserve portion of Great Sand Dunes, the Sangre de Cristo Range rises to Mount Herard (13,297 feet). Such diversity supports a variety of habitats.

There have been forty species of mammals recorded within the park. Among the rodents the Colorado chipmunk and the least chipmunk are common as is the yellow-bellied marmot. In rocky woodlands and brushy areas, the golden-mantled ground squirrel is common. An occasional resident is the Gunnison's prairie dog. It may be sometimes seen along the Park Entrance Road.

Great Sand Dunes squirrels include the pine squirrel, or chickaree, the tassel-eared Abert's squirrel, and the rock squirrel. A common woodland rodent is the porcupine. The woodrat, or pack rat, and the bushy-tailed woodrat are common; the beaver can be found in the wetter areas on the western side of the dunes. Muskrats are seen in wetter habitats.

Perhaps the most unusual rodent in Great Sand Dunes is Ord's kangaroo rat. It is the only mammal that can survive its entire life on the surface of the dune. It avoids predators that visit the dunes at night with its ability to leap five feet into the air. It is able to survive on seeds and grasses found at the edge of the dunes. Its metabolism is so efficient that it does not have to drink.

Lagomorphs include the American pika, Nuttall's cottontail, desert cottontail, and white-tailed jackrabbit. Their habitats are separated by elevation. Raccoons are common in moist areas.

The mustelids (weasels) include the long-tailed weasel, pine marten, and striped skunk. Coyotes are common in Great Sand Dunes. They even wander onto the dunes to hunt rodents.

Black bears are common in Great Sand Dunes, especially in the higher elevations. They may be encountered in the montane forest. A quiet hike on the Mosca Pass Trail may reward the visitor with a glimpse of this predator.

The mountain lion is common but very infrequently seen. The best opportunity to see a mountain lion is on the Mosca Pass Trail. Bobcats may be found in many of the park's environments and even occasionally wander onto the dunes to hunt.

Elk spend summers in the higher elevations of the park and preserve and winter in sandy grasslands. About one thousand elk spend the entire year at the lower elevations. Mule deer are most frequently encountered in brushy and woodland areas. They also are commonly seen in the campground, picnic areas, and parking areas, and along highways.

Pronghorns are occasional to the park. They reside in the grasslands and brushlands of the park. Also occasional is the bighorn sheep. Bands of sheep are resident in the Sangre de Cristo Range.

CONSERVATION CONCERNS

Most visitors visit the park to enjoy the dunes. The evidence of impacts of most activities is quickly obliterated by the wind. Impacts elsewhere in the park could become more evident if visitation grows.

Environmental issues at Great Sand Dunes include potential air and water pollution. As development increases within the San Luis Valley, impacts

The peaks of the Sangre de Cristo Range are seen beyond the dunes, which rise to 750 feet above the surrounding landscape. Prevailing southwesterly winds blow the sand toward the mountains until reverse winds from the peaks stop the movement, resulting in dune creation.

may be felt within the park. Wind-borne pollutants may also be carried from more distant sources. Issues concerning water claims may have an effect on groundwater, ultimately resulting in possible changes in the park's wetlands.

Being downwind from extensive agriculture, exotic plants are a potential problem in the park. Some exotics are already present; vigilance is needed to prevent further infestation.

GREAT SMOKY MOUNTAINS

Misty Valleys and Lofty Peaks

Gatlinburg, Tennessee

ACRES: 521,495

It's 1920 in Elkmont, Tennessee, and the town is bustling. The sounds of train whistles fill the valley. The Little River Railroad train is about to depart for the sawmill in Townsend with newly harvested logs. People have lived here for decades, first attempting to farm the valley and later logging. Eventually the forests were cut out and Elkmont slowly declined. By 1930, people had moved out as the proposed Great Smoky Mountains National Park was discussed. The park was authorized in 1926 and established in 1934, created from private lands. More than six thousand tracts of land were acquired, including most of the town of Elkmont. Return to Elkmont today and you'll find a large campground, usually crowded in summer. The old railroad bed now carries the main park road to Cades Cove.

The park eventually grew to 521,495 acres on both sides of the Tennessee/North Carolina border. The people of Elkmont would never have believed that by 2003 nearly ten million people would visit the park. Nor would they believe the summer traffic or the huge crowds on autumn weekends when the park fills with fall color aficionados.

It's no wonder so many people come to the Smokies. Diversity is the key. Plant and animal species from both south and north meet in the park. The Smokies are the tallest range in the Appalachians with elevations from 875 feet on Abrams Creek to the 6,643-foot summit of Clingman's Dome. Some have estimated that ten thousand species of plants and animals exist in the park.

There are more species of trees in Great Smoky Mountains National Park than in any other national park in the United States or in all of northern Europe. Most of the park is forested, with the trees along the state boundary being part of an old-growth forest. They are left over from the days when logging was too difficult or trees costly to harvest. The humid cli-

mate creates an extended growing season that supports the wide variety of life. The park makes ideal habitat for mammals; there are more than sixty species here.

Visitors want to see black bears. Though they are in "winter sleep" from December to March, bears can be seen almost anywhere in the park. A 2000 estimate of the black bear population was eighteen hundred, a density of more than two bears per square mile.

The bobcat is fairly common. Gray and red foxes are both residents. It is possible that the mountain lion and coyote have moved into the park. Lions were extirpated by 1920, but reports of sightings are becoming more numerous. Their status is still questionable. The coyote has moved east and is now a resident.

Both the striped skunk and eastern spotted skunk are present. Raccoons are found throughout the park at all elevations. Seen along streams is the Virginia opossum. There are two species of cottontails in the park, the eastern cottontail and the Appalachian cottontail.

The river otter was hunted and trapped out of the park by the 1920s. In 1986 river otters were reintroduced and now can be found in both the Tennessee and North Carolina portions of the park.

There are several rodents commonly seen at Great Smoky Mountains. Probably the most frequently seen is the eastern chipmunk. Another commonly seen rodent is the woodchuck or groundhog. The eastern gray squirrel and red squirrel are found in park forests. Other park squirrels include the eastern fox squirrel, southern flying squirrel, and northern flying squirrel. The northern flying squirrel and the Indiana bat are the park's endangered species.

In the 1930s, deer were almost gone from the Smokies due to hunting pressure. After park establishment the population of white-tailed deer grew, and today they are the most common large mammal.

Elk were extirpated from the park by over-hunting and habitat loss. The last elk in North Carolina died in the late 1700s, while in Tennessee, the last one died in the mid-1800s. In 2001, twenty-four elk were reintroduced from a herd at Land Between the Lakes in western Kentucky. Others were added to the herd in subsequent years. This is an experimental reintroduction, and it will be reevaluated in 2006.

A similar experimental reintroduction was attempted recently with the red wolf, the original native wolf in the area. The reintroduction was considered unsuccessful and the wolves removed to facilities elsewhere.

CONSERVATION CONCERNS

In 2004, *National Geographic Traveler* magazine convened a panel of two hundred travel experts to evaluate 115 tourist destinations worldwide. Sites were rated based on environmental conditions, aesthetics, tourism management, and social/cultural integrity. Unfortunately, Great Smoky Mountains National Park is near the bottom of the list. Environmental conditions, aesthetics, and tourism management were all rated as poor. The National Parks Conservation Association (NPCA) concurs. They have placed Great Smoky Mountains National Park on their "Ten Most Endangered Parks" list for both 2003 and 2004.

The nearly ten million visitors to the park are only a part of the problem. Many visitors never leave the park's roads; only a small portion of visitors walk a trail into the park's interior. But ten million people need to be housed and fed. Tourism development crowds the park's boundary. It contributes nearly a billion dollars to the local economy, but slowly the park is being surrounded by commercial and residential developments. Park staff has no control over local development, so the trends continue.

Great Smoky Mountain's place on NPCA's list is based primarily on air quality issues, including aesthetics, acid deposition, habitat damage, and human health. Most of the pollutants come from the burning of fossil fuels outside the park. Sources include power plants (82 percent), factories (13 percent), and motor vehicles (2 percent). Chemicals include sulfates (71 percent), carbon compounds (22 percent), particulates such as soil and dust (5 percent), and nitrates (2 percent). Geographically, the primary sources are the Ohio and Tennessee River valleys and the coast of the Gulf of Mexico.

Visibility in the park continues to decline. Over the past fifty years, visibility has declined 40 percent in winter and 80 percent in summer. Under natural conditions, a vista of ninety-three miles was typical. Today, the average view is only twenty-seven miles, and on the most severe days it drops to less than a mile. Sulfates from electricity generation are the primary source.

Acid deposition, both wet and dry, is also a problem. More oxides of sulfur and nitrogen fall on the Smokies than on any other park that monitors such pollutants. The acids created fall as rain, as an acid cloud, or as dry particles. Average pH of the park's rainfall is 4.5; the clouds sometimes have pH as low as 2.0. Streams are becoming so acidic that the health of the ecosystem is in jeopardy.

Another serious pollutant is ozone, caused by nitrous oxide from factories and motor vehicles.

Research in the park has shown that ozone has seriously harmed much of the park's vegetation. Research further showed that ozone damage is much worse at higher elevations.

A more serious concern of ozone is its impact on human health. Ozone levels in the park are frequently higher than measured in nearby cities like Nashville and Knoxville. Ozone is a well-known respiratory irritant, and concentrations in the park, especially at higher elevations, frequently exceed human health standards. In 1998, the park recorded forty-three days when ozone was considered unhealthy for humans; eleven of those days were consecutive.

In the 1930s and 1940s, the chestnut blight was introduced into the United States, literally destroying the American chestnut. This tree was a key producer of mast for wildlife as well as providing some of the best quality hardwood timber in the east. A major component of the forests of the Smokies was destroyed and has yet to recover. Later, the balsam woolly adelgid was accidentally introduced from Europe. Native European firs are unaffected, but the results were disastrous in the Smokies. Most of the Fraser firs have succumbed, leaving dead snags along the park's roads in the high country and along the nearby Blue Ridge Parkway. The latest scourge is the hemlock woolly adelgid. There are more than eight thousand acres of hemlock in the park, many of them included in virgin stands. Unless a cure is found, another eight thousand acres will be destroyed, leaving only dead snags.

Exotic problems plague the park as well. Plants such as kudzu have found their way in, and control efforts are underway. The European wild boar lives in the park and feeds by rooting just at the soil surface. It destroys habitat used by black bears and even directly consumes some species such as salamanders. The hogs have few if any real predators in the park. To control their numbers, the park has resorted to shooting them by park staff.

Poaching of plants and animals from the park may be a problem. A significant market exists in the Orient for ginseng and black bear parts such as gall bladders and paws. A gallon of bear gall bladder bile is valued at $3,000. In the last ten years, seventy shipments of black bear parts to Asia have been stopped. Do these come from the Smokies?

A hiker climbs a corduroy portion of the Appalachian Trail in the early morning fog. The Smokies are home to a large population of black bears, which hikers generally encounter without incident. The almost ever-present haze gives the mountains their name.

Were the caged bears in the tourist traps near the park born in the park or captured in the park? No one knows the answers to these questions. One estimate is that forty thousand black bears are killed illegally each year in the United States. Unfortunately, many of them probably came from our parks. Until parks have sufficient staff to protect park resources, these kinds of losses will unfortunately continue.

GUADALUPE MOUNTAINS

Rising from the Desert

Salt Flat, Texas

ACRES: 86,416

On the opposite end of the Permian reef that contains Carlsbad Caverns is something totally unexpected. There is the same arid landscape that characterizes Carlsbad's surface, but with some walking you can find ponderosa pine, Douglas-fir, and elk. Guadalupe Mountains National Park in extreme west Texas displays exceptional diversity. Elevation varies from low desert at thirty-six hundred feet up to the highest point in Texas, the summit of Guadalupe Peak (8,749 feet). Drainage from the high country creates canyon environments containing riparian ecosystems. Mountain, desert, and riparian habitats are found in proximity.

This part of west Texas has always been sparsely populated. The Mescalero Apaches ruled the area until the latter part of the nineteenth century. Even so, the Butterfield Overland Mail operated through the future park for one year in 1858–59. Apache attacks forced the route to a more southerly location. Military forays followed that ultimately allowed settlement. Ranchers were not often successful, but some families lasted well into the twentieth century. Much of the former ranch land ultimately became part of the park.

The desert portion of the park is part of the Chihuahuan Desert. Rainfall here is less than twenty inches per year. Plant life includes cacti and other succulents such as agave, sotol, cholla, and yucca. The park's canyons are steep-sided providing sig-

nificant shade for plant and animal life. Many have an abundance of water offering habitats for many mammals. Trees found here include gray oak, velvet ash, and bigtooth maple. At slightly higher elevations, grassland is the primary ecosystem. In the high country there are dense forests of ponderosa pine, southwestern white pine, Douglas-fir, and aspen. This diversity of plant life results in a similar diversity of mammal life.

There are sixty species of mammals in the park. Some live in particular environments; others are widespread in many habitats. The park is relatively small for western parks, 86,416 acres, but the number of mammals here is remarkable. There are few roads in Guadalupe Mountains, so to see many of the park's mammals a hike to the high country might be necessary. Desert mammals must adapt to the heat. One way is to be nocturnal, so spotting some of the desert wildlife may be difficult.

There are a variety of rodents in the park. In the desert, the most common rodents are the Texas antelope squirrel, desert pocket mouse, and Merriam's kangaroo rat. In the grasslands and rocky canyons, the rock squirrel predominates. The gray-footed chipmunk is the only chipmunk in the park. Woodrats include the white-throated, Mexican, and southern plains. Wherever there are trees, you may find the porcupine.

Guadalupe's deserts are home to three species of lagomorphs. Both the black-tailed jackrabbit and the desert cottontail are frequently seen. The rare eastern cottontail is a resident of the high country.

Three species of skunks reside in the park: the western hog-nosed, the striped skunk, and the western spotted skunk. In riparian areas, raccoons are common. The ringtail is widespread in the park. The coyote is seen everywhere in the park as is the gray fox.

Larger predators include black bears and mountain lions. The black bear resides in the park's high country but is uncommon. Park staff conducted a comprehensive study of the mountain lion from 1983–86 to assess its future in the park. Results estimated that the annual population of lions in the park was fifty-eight. Over the course of the study, sixty-five lions were killed around the borders of the park, many related to livestock predation. Nonetheless, research suggests that the lion population in the park will likely remain steady, with reproduction and immigration replacing losses.

The most common ungulate is the mule deer. The only other ungulate you may see is the elk. The elk was an original resident in these mountains but was hunted to extinction in the early twentieth century. It was reintroduced in the 1920s and now may number between fifty and seventy individuals. They are found in the high country, in the canyons, and sometimes in the upper parts of the desert. The collared peccary, or javelina, was introduced in the past and is continuing to hold on.

There are two other ungulates in the park, both exotics introduced in the past. The Barbary sheep, a North African species, lives in the higher elevations. The other, the mouflon from the Mediterranean area, is widespread in the park. Both are sometimes mistaken for the native bighorn sheep, which were extirpated here.

CONSERVATION CONCERNS

Guadalupe Mountains is not heavily visited. In 2003, only 180,000 visits were recorded, less than half of the visitation at nearby Carlsbad Caverns. Access to the park is limited to a few roads that do not penetrate far into the interior of the park. Getting to the higher elevations involves a hike of several hours. Therefore, impacts from visitation are much less than in other parks.

Unfortunately, resource problems can come from afar. Such is the case at Guadalupe Mountains. Air quality was first monitored in 1982. Photos measuring visibility are taken three times per day year round. Data collected in the 1990s showed that visibility on hazy days was getting worse, but better on the clearer days. The poor visibility comes from sources in northern Mexico as well as the southwestern United States. These include sulfates and zinc from metal-smelting activities in Monterrey, Mexico, air pollutants from Los Angeles, sulfates from copper smelting in Arizona and New Mexico, and sulfates from petroleum refining on the Gulf Coast of the United States. Some of the visibility problems are the result of dust and blown soil from local sources. As urban populations continue to grow, air quality will become a greater problem. Air quality legislation provides a basis for controlling domestic sources of pollutants, but international sources will require diplomatic efforts.

The sulfates noted above are the forerunners of acid rain. Acid deposition has been monitored weekly at Guadalupe Mountains since 1984. Rainfall is measured regarding pH, conductivity, and weight. If there has been sufficient rain, further chemical analysis is carried out. Since monitoring began, pH has ranged between 4.30 and 6.70. Acid deposition is not a major concern yet, but as with visibility, the future may bring increases in sulfates.

The impact of humans in the park is not completely understood. Ranching certainly had an effect on vegetation. Native Americans may have used fire

as a management tool, though no one is sure. Today, resource managers are assessing the vegetation in the park to determine if fire can be reintroduced into the ecosystem. The historic role of fire will have to be determined and the presence of fuels quantified before the fire can be used to create the historic vegetative mosaic of the park.

Finally, fire, pollutants, and mountain lions do not necessarily observe boundary lines. Many of the park's problems are the result of activities outside the boundaries. An ethic toward protecting natural ecosystems must be developed. In the case of Guadalupe Mountains National Park, that ethic must have an international component. Only then will long-term protection of park resources be a reality.

The Guadalupe Mountains provide a scenic backdrop to the Chihuahuan Desert vegetation of Pine Springs Canyon. Although the landscape appears dry, the park is home to sixty species of mammals, including coyotes, mountain lions, and black bears.

HALEAKALA
The Volcanic Landscape

Kula, Maui, Hawaii

ACRES: 29,094

Most tourist ads for Hawaii promote beaches, hulas, and flower leis. But there is more to Hawaii than that. Hawaii has some impressive mountains including Maui's 10,023-foot Pu'u'ula'ula Summit within Haleakala National Park. The summit is part of Mount Haleakala, a dor-

The barren vistas of Haleakala National Park tell a tale of volcanic origin. The park is home to Hawaii's only nonmarine wild mammal, the hoary bat.

(Opposite) Cinder cones are reminders of volcanic activity past and present at Hawaii Volcanoes National Park. Lava flowing from Kilauea is one of the park's major attractions.

mant volcano. A paved road provides access to the summit area at the Haleakala Visitor Center. From the visitor center you can look out across the park's 30,183 acres, 24,719 of which are designated wilderness. The view stretches all the way down to the sea. A portion of the park is designated as Kipahulu Valley Biological Reserve, one of the last intact rain forests in Hawaii. Very unusual for national parks, this area is closed to public access.

Haleakala is a hiking park. Roads touch the park in three places, while trails provide access to the wilderness area. Wilderness cabins are also available. More than 1.4 million visitors came to Haleakala in 2003.

The diversity of native wildlife at Haleakala is not as rich as at parks elsewhere. The only native mammal in the park is the hoary bat. In fact, it's the only land mammal in all of Hawaii; the rest are marine mammals.

Historically, birds have been Hawaii's primary wildlife attraction. Unfortunately, an invasion of alien creatures has already driven eighty-five bird species to extinction with thirty-two more listed as endangered on the federal Endangered Species List. Hawaiian ecosystems are surely at risk.

CONSERVATION CONCERNS

If you were to look at a list of mammals resident at Haleakala, you might be surprised. Other than the hoary bat, all are introduced and all are causing havoc with the native Hawaiian ecosystems. Feral animals on the list include dogs, cats, horses, pigs, and goats. Introduced wildlife is present as well: small Indian mongoose, axis deer, and European rabbit. Several rodents are here too: house mouse, Polynesian rat, brown (Norway) rat, and the notorious black rat.

Goats historically grazed in Haleakala's crater resulting in substantial erosion. The pigs were mostly found in the moist parts of park. Between 1976 and 1986, however, park staff built a fence around the boundary of the Summit District. The thirty-four miles of fence is designed to keep the exotic mammals out of the park. In addition, between 1958 and 1993, seventeen thousand goats were removed from the park.

The rats and mice arrived by accident, and to combat them the small Indian mongoose was brought in to prey on rodents in sugarcane fields. But their prey ultimately included ground nesting birds. Hawaii occupies 0.2 percent of the land area of the United States, but one-third of the plants and birds on the federal Endangered Species List are Hawaiian. Aliens keep arriving; Hawaii faces ecological chaos.

HAWAII VOLCANOES
The Fire Mountains

Hilo, Hawaii

ACRES: 323,431

We are periodically awed by nature's fury as displayed by Kilauea. The eruptions of molten magma make spectacular viewing on the nightly news. Most people know Kilauea is in Hawaii, but few realize they are seeing one of the highlights of Hawaii Volcanoes National Park. Nor do they realize the park also contains the highest mountain in the world from base (on the sea floor) to summit, Mauna Loa. At thirty thousand feet, it is one thousand feet taller than Mount Everest.

Hawaii Volcanoes is located on the southern coast of the big island of Hawaii. Its 323,431 acres are arranged in an odd shape to include the summit of Mauna Loa as well as Kilauea's Crater and many of its lava flows. A road leads to the Mauna Loa Observatory at 11,150 feet. A trail climbs the remaining two thousand feet to the summit. The Crater Rim Drive encircles Kilauea Crater and includes the Kilauea Visitor Center sited on the rim.

The park has a frontage on the Pacific Ocean. The Chain of Craters Road leads down to the sea but is now closed due to one of Kilauea's lava flows. Most of the sea frontage is accessible only by trail.

Originally, Hawaii was rich biologically, espe-

mammals is fencing, which is effective against pigs and goats. Where possible, staff tracks and kills exotic animals. Exotic plants are monitored and pulled from the ground. Where exotic flora and fauna are removed, it is hoped the former native inhabitants will return.

Hawaii Volcanoes is a popular park with over a million visits each year. Visitor management techniques are necessary to prevent physical site impacts caused by the presence of humans.

HOT SPRINGS
The American Spa

Hot Springs, Arkansas

ACRES: 5,550

cially in its birdlife. Many of them nested on the ground, safe due to the lack of predators. More than 90 percent of Hawaiian flora and fauna was endemic. One of the most noted is the nene, the Hawaiian goose, which is designated as endangered. There was only one native mammal in the park, the hoary bat, also designated as endangered. Marine mammals have used the Hawaiian Islands especially in winter. The endangered humpback whale is an example.

Early Polynesian immigrants brought exotic mammals with them—dogs, cats, and chickens. These biological colonists would ultimately cause havoc with native Hawaiian wildlife. The Polynesians also brought new plants to cultivate including taro, sweet potato, coconut, banana, and sugarcane. Many of these were cultivated on an enormous scale, destroying much of the native vegetation.

CONSERVATION CONCERNS

The biggest concern in Hawaii is the destruction of native environments by exotic plants and animals. The descendants of the domestic dogs, cats, and pigs brought by the Polynesians and later immigrants became feral. They have been joined by horses and goats. Several rodents arrived as well—the house mouse, Polynesian rat, Norway rat, and black rat. In order to combat the rodents, especially in sugarcane fields, the small Indian mongoose was introduced. It found ground-nesting birds to be easy prey.

One of the most important management objectives in the park is to protect what is left of the park's natural environment. Some parts of the park have been damaged beyond repair; management focuses on those places where rehabilitation is possible. One method used in dealing with the exotic

Most national park aficionados know that Yellowstone National Park was the world's first, established in 1872. There might be an argument from Arkansas residents, however. The Hot Springs of that state were known early in the nineteenth century, and by the 1830s, cabins and a store had been constructed. Responding to public pressure, President Andrew Jackson withdrew the lands around the Hot Springs, reserving them for future disposal. This was the first time the federal government set aside lands that would eventually become a national park.

Though private citizens tried to claim the springs, they remained in federal ownership, and in 1921 Hot Springs was declared a national park. Hot Springs National Park has been a federal reserve forty years longer than Yellowstone, but official designation came nearly fifty years later than Yellowstone.

Even though the springs remained in federal ownership, a spa-focused city grew up around them. Eventually, Hot Springs became known as "America's greatest health resort." Elegant buildings and landscaping flourished, especially the noted Bathhouse Row. National park designation resulted in the construction of eight new state-of-the-art bathhouses. Promenades, mountain drives, and fountains were added as well. Tourist hotels sprang up on nearby private land.

Eventually, visits to spas declined as a tourist

attraction. By 1985, only the Buckstaff Hotel remained open for business. The Fordyce was remodeled as the park's visitor center and museum. The others remain vacant awaiting possible adaptive uses.

Many of Hot Springs National Park's 5,550 acres are adjacent to Hot Springs, a city of thirty-three thousand residents. Mountainous park land also surrounds the city, especially on its west, north, and east sides. The park is primarily oak/hickory/pine forest, typical of the south central United States. Given the proximity of the city, the mammals present are primarily those that tolerate humans.

Eastern gray squirrels live throughout the park's forests. The eastern chipmunk is found everywhere in the park. The fox squirrel is uncommon. Woodchucks are infrequently seen in the park. The southern flying squirrel is nocturnal and seldom seen.

The most abundant mammal in the park is the eastern cottontail. Most of the other common mammals come out at night. These include the Virginia opossum, raccoon, nine-banded armadillo, and striped skunk. A few white-tailed deer reside in the more hilly parts of the park.

The park's only observed carnivores are the foxes. The gray fox hunts at night; the red fox is extremely rare. The bobcat and coyote are probably present but rarely observed.

CONSERVATION CONCERNS

Resource issues at Hot Springs National Park are related to the presence of people. Having the city interspersed with park lands causes many problems. Motor vehicles cause air pollution as well as traffic congestion. Mammals and other wildlife crossing highways are at risk.

Another urban problem is trash dumping on park lands. In addition, where city residents aren't careful about garbage, park mammals become trash raiders. Skunks and raccoons are frequently the culprits.

Exotic plants and animals can also be a problem. Residential plantings or exotic pets may end up competing with resident wildlife. Feral dogs and cats are found in the park as are house mice and Norway and black rats.

For a small park, Hot Springs attracts considerable visitation. In 2003, more than 1.5 million people visited the park. With them come traffic congestion, trash, and air pollution. Park staff are in the midst of conducting resource inventories so that trends in wildlife population and health are well documented.

ISLE ROYALE
The Wild North Woods

Houghton, Michigan
ACRES: 571,790

When we think of summer crowds in our nation's parks, we think of the millions of visitors to Great Smoky Mountains, Yellowstone, and Yosemite National Parks and the related traffic. There are, however, parks where no such crowds exist. In fact, at Isle Royale National Park, annual visitation is less than a busy day at one of the parks noted above. In 2003, fewer than eighteen thousand people came to Isle Royale.

Located in Lake Superior, Isle Royale is harder to visit than most parks. Access is mostly by boat, though a few people arrive by floatplane. Passenger vessels are operated by the National Park Service from Houghton, Michigan, and by private operators from Copper Harbor, Michigan, and Grand Portage, Minnesota. Some visitors arrive by means of private recreational watercraft.

There are no roads or vehicles on the island. Instead, 165 miles of trail provide access across the forty-five-mile length of the island. Lodging and food service are available at Rock Harbor Lodge where there is also a marina and a small store. The National Park Service operates a visitor center near the ferry dock. Elsewhere on the island, thirty-six backcountry campgrounds are available for backpackers and recreational boaters.

Isle Royale contains 571,790 acres; however, much of it is the waters of Lake Superior. Almost the entire island (132,018 acres) was designated wilderness in 1976.

Isle Royale is mostly forested, including spruce-fir and birch-maple forest types. Parts of the island are still in shrubs from past forest fires. Considerable riparian habitat exists, much of it created by the island's industrious beavers.

Isle Royale is a classic example of island biogeography. There are only twenty species of mammals on the island, while on the opposite shores of Lake Superior as many as forty species may be found. In the winter of 1948 the surface of Lake Superior froze, allowing certain species to cross over to Isle Royale.

This event caused long-term change in the island's mammalian fauna.

The gray wolf and the moose are the premier mammals on the island. The gray wolf arrived from Canada by walking across the lake ice in 1948. The next year a wolf study was initiated that continues today. The wolf and its prey species are monitored annually, especially over winter when they are most visible.

Data for 2003 showed that nineteen wolves in three packs inhabited Isle Royale. That was an increase of two over 2002. Wolf populations in the 1970s were much larger, between fifty and sixty individuals. The population plummeted by half in the early 1980s and has remained around twenty to twenty-five since. It is believed that the entire population is descended from one female, resulting in inbreeding that has led to loss of genetic variability.

Isle Royale's wolves range throughout the island but are seldom seen by visitors. They are extremely secretive and avoid humans. Hearing a howl or seeing a footprint is probably the best you can hope for.

The wolf's primary prey is the moose. There were no moose on the island in the early twentieth century. Just when and how they arrived is unknown. They either crossed the ice when the lake froze in 1912 or swam from Canada. When wolves arrived, the populations of moose and wolf became intertwined. The 2003 moose count showed nine hundred individuals.

Moose are abundant and commonly seen on land and eating aquatic vegetation in ponds and streams. There tend to be more moose on the ends of the island and fewer in the middle. The severity of winter is a factor in determining moose distribution.

The red fox is common on the island. However, in the past few years its population has declined considerably. It still is frequently seen at backcountry campsites and sometimes around the docks at Rock Harbor.

Several members of the weasel family reside on the island, though none is common. They include the striped skunk and river otter.

The population of snowshoe hares follows a cyclic pattern. For 2002, the population was down after a decade of increase. Beaver populations also vary. They are found in inland lakes as well as the ponds they create.

The most commonly seen mammal is the red squirrel. It feeds on the seeds of the park's conifers so is most often seen in the spruce-fir forest.

Several mammal species used to be present and now do not exist. Most disappeared before the arrival of the wolf. Missing species include the coyote, marten, woodland caribou, and white-tailed deer. No effort has been made to reintroduce these mammals. The lynx was also a member of the above list; its last recorded sighting was in 1981.

CONSERVATION CONCERNS

Located in the middle of Lake Superior, Isle Royale avoids most of the problems associated with high use or activities of landowners on park boundaries. Unfortunately, location does not prevent air quality problems. Pollutants, including toxic chemicals, travel long distances by air currents. Air quality monitoring equipment was added to the Mt. Ojibway fire tower in 1987 to measure particulates, sulfur dioxide, and ozone. Visibility is measured once per day during winter and three times per day during summer.

Air pollution causes several problems including loss of aesthetics, harm to plants and animals, global warming, acid deposition, and accumulation of toxic chemicals. To date, these problems remain a concern for the future. Air pollution is present but is considered low at this time. However, a study of northern pike in some of the island's in-

A cow moose grazes in the morning mist on Isle Royale National Park. Absent at the turn of the nineteenth century, moose somehow made their way to the park from Canada. The Isle's twenty to twenty-five wolves prey primarily on the moose.

land lakes found that they contain mercury above standards. Further research will be conducted to assess the situation.

Water pollution does not appear to be a problem at present. Lake Superior is gigantic and has few polluting industries on its shores. Its clean waters are a major factor in the Isle Royale experience.

JOSHUA TREE
Natural Gardens of the Mojave

*(Pages 56–57)
Mountain goats
travel along the edge
of a glacier in Kenai
Fjords National Park.*

*A member of the
yucca family, the
Joshua tree is the
symbol of this
southern California
national park. With
its blend of different
types of deserts,
the park has high
biodiversity.*

Twentynine Palms, California
ACRES: 789,745

Considering the name, one would think that Joshua Tree National Park was created to protect America's tallest yucca. A member of the lily family, the Joshua tree is native to California and Arizona. It can grow to a height of thirty feet. In reality, Joshua Tree is much more. The park is a unique combination of three ecosystems. In the northern part of the park is found the Mojave Desert, home to most of the Joshua trees. The southern part of the park is part of the Colorado Desert, an extension of the Sonoran Desert into California, and the western part of the park is the Little San Bernardino Mountains. These reach an elevation of four thousand feet, tall enough for pinyon pine and California juniper to exist. The result is considerable diversity. The desert here is brimming with life, including fifty-two species of mammals.

Even with the harsh environment, humans have lived here for five thousand years. Several successive native cultures used the area for hunting and gathering. After the passage of the Homestead Act, families moved into the area, trying to make a living from the land. A few made it; most didn't. Miners followed. Some prospered, like the Lost Horse Mine, which extracted 10,000 ounces of gold and 16,000 ounces of silver.

In the 1920s, roads allowed access to land developers as well as plant poachers. A Pasadena resident, Minerva Hoyt, began a quest to protect the area, especially from plant loss. The result was the creation of Joshua Tree National Monument in 1936. In 1994 Joshua Tree was upgraded to national park status. Today, the park contains 789,745 acres, 585,000 of which are designated wilderness.

In order to cope with heat and lack of water, the mammals of Joshua Tree tend to be small. Of the fifty-two species, fifteen are bats and twenty-four are rodents. Many of the rodents are common in the park. Perhaps the most widespread of them is the white-tailed antelope squirrel. The Merriam's chipmunk, Mojave ground squirrel, and round-tailed ground squirrel are commonly found. Three species of woodrat inhabit the park including the desert woodrat, or pack rat, the eastern dusky-footed woodrat, and the white-throated woodrat.

Other small mammals are present. The park's two species of lagomorphs are the desert cottontail and the black-tailed jackrabbit. Members of the park's weasel family include the western spotted skunk, striped skunk, badger, and long-tailed weasel.

Canid predators are common at Joshua Tree. The coyote is common throughout the park. The gray fox is more nocturnal and less often seen. The desert kit fox is common in the park, at least locally.

A common feline predator is the bobcat. It is nocturnal and highly secretive. Its larger kin, the mountain lion, is uncommon. The black bear may be an occasional migrant through the park.

Mule deer are the most common ungulate in the park. They are found most often in the pinyon-juniper forests at higher elevations. The other park ungulate is the desert bighorn. There are about 250 of them in the park in three major herds. The smallest herd is in the Wonderland of Rocks; these sheep are the most often seen.

CONSERVATION CONCERNS

Joshua Tree has one thing in common with Great Smoky Mountains, Everglades, and Yellowstone National Parks—all are on the 2004 "Ten Most Endangered Parks" list of the National Parks Conservation Association (NPCA). There are several reasons for the listing: air pollution, encroaching development, and exotic species.

Air quality problems are several. Air quality has been monitored in the park since 1978, and data sometimes tell a sad story. On some days, Joshua Tree has the dirtiest air, in terms of visibility, of any park in the country. Most of the haze is caused by smog from surrounding areas. The Los Angeles basin with its huge population is just to the west. New power plant construction nearby also contributes to pollution. Summer is the worst time for pollution; in winter, prevailing winds take Los Angeles air away from the park.

Also during summer, ozone standards are routinely exceeded and, like visibility problems, are frequently the highest of any park. Damage by ozone to the park's vegetation has been documented. Environmental Protection Agency standards regarding human health are frequently exceeded as well.

Growth in the Coachella Valley, which contains Palm Springs and Indio, continues. Development inches toward the southern boundary of the park. A proposal to build a city on the park's boundary has been stopped for now, but private landowners still hope to develop their property. The impact on the park would be considerable. A proposal to develop the Eagle Mountain Landfill to the southeast of the park is still being considered. It would be one of the largest ever developed in the United States.

The presence of exotic plants is a concern and is related to air pollution. Nitrates are deposited on nitrogen-poor soils, which allow exotic plants to flourish. These exotics carry fire more readily than native plants, increasing the possibility of fire damage.

KATMAI
Valley of Ten Thousand Smokes

King Salmon, Alaska
ACRES: 4,725,188

Americans commemorate the Allied landing in Normandy on D-Day, June 6, 1944. But for a few people, June 6 has a different meaning. On that date in 1912, the largest volcanic explosion of the twentieth century happened in Alaska. In a place that would one day be Katmai National Park, earthquakes signaled the coming eruption. The explosion was heard throughout Alaska; the ash fall covered much of the state and as far away as Puget Sound. At the end of the sixty-hour eruption, more than eight cubic miles of material had been ejected into the atmosphere. Surprisingly, no one was killed or injured.

The next year Dr. Robert Griggs of the National Geographic Society visited Katmai to study the volcanic activity and returned several times thereafter. Griggs felt the area was extraordinary and pushed for federal protection of Katmai. In 1918, Katmai was declared a national monument, and in 1980 it was upgraded to national park status. The park now contains 4,725,188 acres, more than three million of which have been designated as wilderness.

Development of Katmai has been minimal. Access is still primarily by air. In 1950, the air tourism industry established Brooks Camp and Grosvenor Lake Lodge, providing the first tourist facilities in the park. A twenty-three-mile dirt road was constructed from Brooks Camp to the Valley of Ten Thousand Smokes; it remains the only road in the park (except for a short segment at Lake Camp).

Visitation at Katmai has declined recently. In 2003, there were 23,754 visits, a decline from 67,306 in 2001 and 59,266 in 2002. The worldwide decline in tourism certainly plays a role, but the difficulty of access and the remoteness of Katmai are also factors.

Katmai National Park contains some of the most spectacular scenery in Alaska: mountains, glaciers, and a wild sea coast. But another major attraction of the park is the largest concentration of Alaskan brown bears in North America. Alaskan brown bears live throughout the park, especially in the coastal areas and in the vicinity of the lakes. The rich protein diet from fish allows some bears to grow to nine hundred pounds. The best place to see these giants is at Brooks Camp where a viewing platform has been built.

Keeping visitors safe in bear country is crucial. In 2003, two visitors were killed by bears in the coastal portion of the park. There is a regulation requiring visitors to remain at least fifty yards from any bear, the violation of which was likely a factor in these deaths. The size and strength of these animals is awesome; visitors must heed park rules to ensure a safe experience.

The other large mammal commonly seen is the moose. They are common around lakes and frequently seen feeding on aquatic vegetation. The park's other large ungulate is the caribou, which spends a portion of the year in the park. The Alaska Peninsula herd, which contains as many as fifteen thousand animals, ranges throughout the Katmai region.

The gray wolf is common and sometimes shows up at Brooks Camp, but otherwise it is seldom seen except by fortunate backpackers. The wolverine occurs throughout the park. Occasionally it is seen along the road to the Valley of Ten Thousand Smokes.

Both the snowshoe hare and the lynx are common in the park. Chances of seeing a lynx improve in the summer when they are forced to hunt in daylight since days are so long. The Alaskan hare exists in the park but is rarely seen.

Smaller mammals abound in Katmai. The hoary marmot lives in dens below talus slopes. The Arctic ground squirrel is common in valley areas. Alaska's largest tree squirrel, the red squirrel, is common in forested areas. Beaver live in many of the park's lakes. Porcupines are common in forested areas; they are nocturnal but are frequently seen around Brooks Camp.

A bear surveys his lakeshore domain in Alaska's Katmai National Park. The park was the site of an enormous volcanic eruption on June 6, 1912.

There are numerous small mammals at Katmai including lemmings, voles, and mice. The small predators include marten, short-tailed weasel, least weasel, river otter, and mink.

Marine mammals are found along the park's rugged coastline. The sea otter is abundant along the coast of the Alaska Peninsula. Katmai is also home to northern sea lions (also called Steller sea lions), a threatened species whose numbers are of concern. The more numerous harbor seals congregate along the shore. Offshore, several species of whales may be seen during migrations. These include the blue whale, gray whale, and killer whale or orca. The beluga whale occasionally swims up tributaries to feed on salmon smolts migrating to the sea.

CONSERVATION CONCERNS

Remoteness is always a positive factor when considering recreation resource issues. Katmai gets few visitors, so direct human impacts are small, though not absent. Visitor deaths by bears in 2003 reinforce the need for visitor management and interpretive programs for bear safety.

Water quality in the park remains essentially unaltered by human actions. The same is true for air pollution though natural sources of ash will periodically fall on the park.

In 1989, Katmai was not remote enough to avoid environmental disaster. That year, the tanker *Exxon Valdez* went aground four hundred miles away, dumping eleven million gallons of crude oil into Prince William Sound. Sea currents carried the oil all the way to Katmai's coast, fouling it with an oily sludge. Wildlife was directly harmed before the cleanup could be carried out.

One issue not present at Katmai is subsistence. Katmai is one of Alaska's parks where subsistence hunting and fishing are not authorized.

KENAI FJORDS
Landscape of the Alaskan Peninsula

Seward, Alaska
ACRES: 669,983

The late 1970s were a critical time in the establishment of national parks in Alaska. Congress and several administrations debated how federal lands in Alaska should be allocated to national parks, forests, and wildlife refuges, or perhaps for some other purpose. In order to prevent development of some of the more special places, in 1978 President Jimmy Carter declared several of them to be national monuments. With passage of the Alaska National Interest Lands Conservation Act in 1980, some monuments were upgraded to national park status. One such place was Kenai Fjords National Park, containing 669,983 acres.

Kenai Fjords is an incredible combination of sea, rock, and ice. The Harding Icefield covers more than half of the park and receives more than four hundred inches of snow annually. From the icefield, thirty-two named glaciers flow to the sea. Where glaciers have retreated from the sea, fjords were formed with steep sides and deep waters. Where the ice has retreated, forests are beginning to form. The succession from bare soil to forest is especially visible at Exit Glacier.

One of the differences between Kenai Fjords and other Alaskan parks is that it is accessible by highway. The park's visitor center is in the town of Seward. It is only a short drive to Exit Glacier, where you can walk up to the face of the glacier, a glacier that has retreated and advanced in recent years. To see the fjords and remote parts of the park, boat trips are available daily from Seward harbor. Air charters are also available to the fjords and even onto the icefield. In 2003, 236,940 visits were recorded at Kenai Fjords.

Even though Kenai Fjords has more than six hundred thousand acres, much of it is ice covered and consequently not inhabited by mammals. Even so, thirty species of mammals live in the park, both on land and in the sea.

Both the black and brown bear are found in Kenai Fjords. The larger Alaskan brown bear lives near Exit Glacier and Resurrection Bay, places where the greatest number of people are found. The campground at Exit Glacier has been specially designed to protect visitors from potential bear encounters.

The Exit Glacier area provides excellent habitat for moose. Mountain goats are often seen on the cliffs above Exit Glacier or on the rocks above some of the fjords. They look like white specks high above. Goats can be confused with Dall's sheep, which only infrequently enter the park.

The coyote is widespread but avoids areas where the wolf is common. The gray wolf is present but is highly secretive and seldom seen. The lynx and its favorite prey, the snowshoe hare, are found in the park.

Several members of the weasel family are present; wolverine, river otter, American marten, mink, and ermine (also called short-tailed weasel) are residents.

The hoary marmot is found in rocky areas. The red squirrel and northern flying squirrel are Alaska's only tree squirrels. Beaver live in riparian areas while the porcupine is common in the forests.

Marine mammals are commonly seen in the park. The sea otter is common in Resurrection Bay and elsewhere. It is sometimes observed off the beach in Seward near the park visitor center.

The others are more often seen while on one of the park's boat tours. Two pinnipeds are common. The northern sea lion (Steller sea lion) is often seen at the haul-out site on the Chiswell Islands. The other pinniped is the harbor seal.

Two species of porpoise live in park waters. Dall's porpoise are usually found in groups of up to twenty and follow tour boats. Conversely, the harbor porpoise is very shy and avoids larger vessels.

Whales are both resident and migrants in the park. The killer whale, or orca, lives in the park. The other whales are migrants through the park and may be seen from the tour boats. Whales that may be encountered include the bowhead, Minke, sei, blue, fin, humpback, and gray.

CONSERVATION CONCERNS

Resurrection Bay was spared oily shores from the *Exxon Valdez* disaster of 1989. There is no guarantee, however, that the same will be true if another oil spill should occur. Seward is becoming a more important port for cruise ships and other large vessels. Crude oil is not exported from Seward. Years ago, a considerable tonnage of coal was shipped to Asia through Seward. Accidents do happen; how we respond to them will determine how serious oil spills in the park might be.

Kenai Fjords received nearly a quarter million visitors in 2003. However, this use is concentrated

Bald eagles perch on a snag while snow falls in Kenai Fjords National Park. Unlike most of Alaska's national parks, Kenai Fjords is accessible by automobile. One can literally drive up to a glacier.

during the summer season and in the Exit Glacier area. With the cruising industry in an expansive mode, it is very possible that congestion and impact to the park may increase.

So far, air and water pollution have not been problems and are not likely to become so in the short run. Long term, however, new industry, more cruise ships, and more people all could contribute to a decline in environmental quality.

KOBUK VALLEY
River Bluffs of Frozen Sand

Kotzebue, Alaska

ACRES: 1,700,000

It is difficult to say which national park is the most remote. Dry Tortugas, Isle Royale, and Channel Islands are all islands, but some Alaskan parks are probably more difficult to get to. Lake Clark, Glacier Bay, Gates of the Arctic, Katmai, and Kobuk Valley are not on any highway; Kobuk Valley is probably farthest from any road. The 1.7 million-acre park is 350 miles northwest of Fairbanks and seventy-five miles east of Kotzebue, location of the nearest road. The nearest towns to Kobuk Valley are Ambler (pop. 262) and Shungnak (pop. 238); both have scheduled air service from Kotzebue. Both are on the Kobuk River where boating or rafting into the park is possible. Landing of fixed wing aircraft is allowed throughout the park; landing of helicopters is not.

Kobuk Valley National Park is located just south of Noatak National Preserve (6.5 million acres) and just west of Gates of the Arctic National Park and Preserve (eight million acres). These contiguous areas total more than sixteen million acres of Alaskan wilderness. There are no developments in Kobuk Valley, or roads, trails, campgrounds, or picnic areas. Visitors must come self-sufficient.

Kobuk Valley is a semi-enclosed mountain basin with the Kobuk River flowing through the valley. South of the river are the Great Kobuk Sand Dunes, a popular attraction for park visitors. The valley bottoms have birch, willow, and balsam poplar; spruce forests are also evident. Tundra ecosystems are common in the park.

Kobuk Valley is home to Alaska's "Big Five"— grizzly bear, wolf, Dall's sheep, caribou, and moose. The nomadic caribou numbers in the thousands. Their migration through the park begins in March. The caribou pass through the park on their way to their calving grounds on the Arctic Coastal Plain. Their post-calving area is to the west, north of Cape Krusenstern. They move south, crossing the Noatak in August and the Kobuk in September. Their wintering grounds are south of the Kobuk River, immediately south of Kobuk Valley National Park.

Both grizzly and black bears live in the park. Grizzlies at Kobuk Valley prey on caribou and other animals, but their diet is omnivorous with grasses, sedges, and berries a significant portion of their food. The black bear is an animal of the forest; they too are omnivores. Black bears are usually scarce in areas where grizzlies are common.

Considerable evidence of the presence of wolves is found throughout the park. Wolves follow the caribou herds, taking the old, sick, and lame, as well as the young. Some wolves remain in one area and hunt the caribou herd as it passes. Similarly, the red fox inhabits wide areas of the park. They too hunt caribou but are only able to take calves.

Some of the smaller predators can also be found in Kobuk Valley. The wolverine is not common anywhere in its range, but it occurs along the Kobuk River. The American mink has also been documented in the park, but its population is extremely limited. The least weasel lives in a variety of habitats wherever rodents are common but is very uncommon in the park.

Moose are common along the Kobuk River and its tributaries. Users of the river will have a good opportunity to see them. Dall's sheep are uncommon; they are erratically distributed in the northern half of the park.

Kobuk Valley is home to the porcupine. They prefer forested habitats but are quite rare. Surprisingly, a specimen was collected in the Great Kobuk Sand Dunes.

The rodents are the most numerous group in the park and provide the basis for the food chain for many of the predators. The Arctic ground squirrel is common throughout the park. In the spruce forests, the red squirrel is common. Several species of voles and lemmings are also common.

CONSERVATION CONCERNS

Kobuk Valley National Park receives very little visitation. In 2003, 4,006 visits were recorded, a slight decline from the previous two years. Even so, there are some sites affected by humans, such as put-in points along the river and popular campsites. There is, however, no air pollution from vehicles. In addition, Kobuk Valley's remoteness reduces the probability of air and water pollution being carried into the park. The Kobuk River has no development upstream from the park except for several bush villages. Some sewage spillage is a possibility.

Air pollution is not a problem as yet but could become one in the future. If there is energy development in Alaska, will there be further pipeline building and if so, where will it be?

Finally, Kobuk Valley is one Alaskan park that allows subsistence use for hunting and fishing within park boundaries. There have been instances in the past where subsistence activities, especially fishing, have been interfered with by recreational visitors. Subsistence is a lifestyle for bush people, and such interference may cause hardship for these families.

LAKE CLARK
Coastal Cliffs and Mountain Glaciers

Port Alsworth, Alaska

ACRES: 4,030,025

The Alaska National Interest Lands Conservation Act of 1980 (ANILCA) was designed to allocate 104 million acres of federal lands in Alaska to national forests, wildlife refuges, and parks, as well as to the Bureau of Land Management. When President Jimmy Carter signed the law, fifty million acres were added to the National Park Service, creating ten new national park areas. One of these was Lake Clark National Park and Preserve.

There are no roads to Lake Clark. Access is most often by small aircraft from Anchorage or one of the towns on the Kenai Peninsula. There are commercial establishments in the park, including outfitters and lodging. Visitation at Lake Clark is quite low, about forty-five hundred visits annually for both 2002 and 2003.

Lake Clark contains four million acres of some of Alaska's finest scenery. Two active volcanoes, Mt. Redoubt (10,197 feet) and Mount Iliamna (10,016 feet), are the most visible of the park's mountain peaks. They can be seen from across Cook Inlet by travelers on the Kenai Peninsula. In the park, peaks of the Alaskan and Aleutian ranges join, creating an extraordinary panorama of rocks and ice. Several mountains are glacier covered; western slopes are characterized by tundra vegetation. Finally, the park has frontage on the west side of Cook Inlet.

Lake Clark also has deep valleys. Several contain lakes including the forty-mile-long Lake Clark. Three wild rivers flow through the park, the Tlikakila, Mulchatna, and Chilikadrotna. All three offer spectacular floating opportunities. One of the

major reasons the park was created was to protect and enhance the fishery of nearby Bristol Bay.

The brown bear is found throughout the park. They are most common in the coastal area along Cook Inlet. They're also often seen around Lake Clark, especially near Kijik River where there is a major salmon run. Black bears are locally abundant. Wolves are usually found in mountainous areas in forests below the rocks and ice. They also live in tundra areas.

Caribou spend a majority of their time in the tundra. The area around the lakes in the northwestern part of the park is used by caribou for a part of the year including calving. The herd is estimated at two hundred thousand animals and growing.

Moose are found everywhere in the park below timberline. Moose typically feed in or near bodies of water, especially on aquatic vegetation. Dall's sheep live at high elevations. There are approximately six hundred of the all-white sheep in the park.

Red foxes are quite common and can be found in all types of landscapes. Coyotes can be found in the park's brushy and grassy areas. Lynx are also found throughout the park at most elevations. Members of the weasel family are common in the park. The marten, wolverine, short-tail weasel (ermine), least weasel, and river otter are park residents.

There are numerous species of rodents in the park. Beaver are found everywhere. Porcupines are very common in forested areas. The hoary marmot is usually found in the upper alpine areas. The Arctic ground squirrel is an extremely important food source for the larger predators including wolverines, wolves, and bears.

Two species of lagomorphs are found at Lake Clark. The snowshoe hare is common throughout the park except near the coast. In rocky areas in the alpine zone, the collared pika is common.

The Cook Inlet coast sometimes provides an opportunity to see marine mammals. The coast freezes during winter, so there is no year-round population of marine mammals. The harbor seal is quite common in season; the northern sea lion (Steller sea lion) less so.

Whales are sometimes seen in the area. Minke and humpback whales are migratory and resident seasonally. Killer whales and white beluga whales may be seen more frequently.

With no roads in Lake Clark National Park, visitation is rare. Over the course of an entire year, fewer than five thousand people visit this Alaskan park. Fishermen who fly into the park often share the space with brown bears.

CONSERVATION CONCERNS

Both air pollution and water pollution are being monitored in the park. Particulate matter is commonly spewed into the air by the park's volcanoes. Otherwise, air pollution is not a problem as yet. Park waters currently surpass all state water quality standards. There are no activities in the park's watershed that would contribute pollutants to park waters. A concern is potential growth in visitor numbers as well as activities on private lands within the park. Much of the shoreline of Lake Clark itself is in private hands. The village of Port Alsworth (pop. 40) is a privately owned village on the lakeshore where sewage is handled by septic tanks. Wave action from power boat wakes may also cause erosion and subsequent siltation.

Airborne deposition is not a problem at present, but changes in industrial development worldwide could allow pollutants to enter the park in the future. Monitoring such changes is a key first step in preventing environmental damage.

Lake Clark National Park was authorized by ANILCA to allow subsistence hunting and fishing. Conflict with sport fishing in the park may occur. Subsistence is an Alaskan issue whose outcome is yet to be fully understood.

LASSEN VOLCANIC
Peaceful Forests above the Restless Earth

Mineral, California

ACRES: 106,372

When Washington's Mount St. Helens erupted in 1980, news people began looking back to see how long it had been since a volcano erupted in the lower forty-eight states. Such eruptions had occurred in Alaska and Hawaii, but there had not been active volcanoes in any other state since 1914. That year Lassen Peak began erupting and continued for seven years. In 1915 the Great Hot Blast threw ash five miles into the air while a part was deflected down a massive mudslide from the previous year. Now called the Devastated Area, it has demonstrated how a landscape recovers from volcanic activity. The Lassen Peak area had

been a national monument since 1907, but after the eruption the area was designated a national park on August 9, 1916, to protect the mountain and the surrounding volcanic evidence.

Lassen Peak is located on the southern end of the Cascade Range at the point where the Cascades and Sierras meet. The northeastern portion of the park is a part of the Great Basin. The Cascades extend north into Canada and are part of the Pacific Ring of Fire. Will it erupt again? No one knows for sure. The hot springs at Bumpass Hell and Sulfur Works suggest that Mother Nature is still at work.

The vegetation of Lassen Volcanic National Park is typical of the Cascade Range, though some species from the Sierras to the south are also found in the park. Coniferous forests are the most common; some are old growth. Aspens and willows are also present. Elevation varies from five thousand feet up to the summit of Lassen Peak at 10,457 feet. These habitats support fifty-six species of mammals.

Black bears are uncommon and usually found only in the backcountry. Mountain lions live in the same area. The smaller predators include the canid family; the coyote is probably the most common. Gray foxes reside in lower elevations.

Several members of the weasel family live in the park; none is common. These include the marten, mink, river otter, badger, striped skunk, western spotted skunk, short-tailed weasel (ermine), and long-tailed weasel.

Porcupines live in the forests of the lower elevations; raccoons also frequent the lower elevations. The Virginia opossum was introduced to the park and lives near the park's boundaries at the lower elevations.

Several species of lagomorphs exist in the park. The American pika lives in rocky areas at high elevations. Conversely, mountain (or Nuttall's) cottontails are abundant at low elevation but seldom seen. Snowshoe hares are found in forested areas. The black-tailed jackrabbit is present in the park.

The most frequently seen among the rodents are the golden-mantled ground squirrel and the yellow-bellied marmot. Douglas's squirrel, or chickaree, is common in the forested areas of the park. The northern flying squirrel is present but rarely seen. Other squirrels include the western gray squirrel, California ground squirrel, and Belding's ground squirrel. Several chipmunks inhabit the park: lodgepole chipmunk, yellow-pine chipmunk, least chipmunk, long-eared chipmunk, and Merriam's chipmunk. In rock-slide areas, the bushy-tailed woodrat resides.

The primary ungulate in the park is the black-tailed deer, an abundant subspecies of the mule deer. It lives at low elevations outside the park during the winter and migrates to higher elevations during spring to take advantage of new vegetation.

CONSERVATION CONCERNS

Lassen Volcanic National Park is located away from the major population centers of California. Visitation is comparatively low—406,782 visits in 2003. Most of the park's roads are closed in winter except for access to the Lassen Winter Sports Area at the southwest entrance of the park. Some parts of the park, therefore, get significant use during the summer.

Park staff has established a monitoring program to assess the effects of park visitation on water quality and the quality of water on park watersheds. So far, no significant problems exist. Some of the park's lakes have been dammed, and some have had exotic fish stocked in them. Restoring naturalness may be considered in the future. A catch-and-release policy may be established for fishing park lakes.

Not only has Lassen been significantly changed by volcanic action, human-caused changes have also occurred. Parts of the park have been heavily grazed in the past by horses, sheep, and cattle. Most wildfires have been suppressed over the past ninety years. Rehabilitation will likely be necessary in order to allow natural succession to proceed.

Lassen Volcanic National Park is approximately two hundred miles north of the Sacramento area, a source of air pollutants. At times, visibility is degraded at Lassen by pollutants from vehicles and industry. Ozone is often a problem as well. Even though the Clean Air Act requires the enforcement of national ambient air quality standards, identifying and convicting polluters is sometimes difficult.

MAMMOTH CAVE
The World's Longest Cave System

Mammoth Cave, Kentucky
ACRES: 52,830

Mammoth Cave has long been known by humans. There were Paleo-Indians in the area ten thousand years ago. Archaic and Early Woodland peoples actually entered the cave four thousand years ago to collect minerals such as gypsum from cave walls. They left evidence of their visits, including cane reed torches, sandals, and even mummified bodies.

Europeans moved into the region in the late 1770s. John Houchin settled on the banks of the Green River just downstream from Mammoth Cave. He is credited with discovering the cave while chasing a bear, though the story is somewhat questionable.

The cave was well known by the early 1800s, and during the War of 1812 Mammoth Cave was mined as a source of saltpeter. Huge amounts were shipped east for manufacture into gunpowder until 1815. In late 1815, Nahum Ward visited the cave. Soon after, he made a new map of the cave and penned a long description that ultimately appeared in books, newspapers, and magazines of the day. This notoriety attracted travelers, and soon Mammoth Cave was open for tours.

Exploration continued; tours were provided by slaves of the cave's owner. One slave, Steven Bishop, became the cave's best known explorer and guide. For the next hundred years, Mammoth Cave was a major tourist attraction. Various landowners opened cave sections and competed for business.

In 1912, Congress held its first hearings on establishing a Mammoth Cave National Park. Like other parks in the east, national parks could only be created by donation to the federal government. In 1924, the private Mammoth Cave National Park Association was created to promote the park. They were joined by the Kentucky National Park Commission in acquiring park lands. In 1941, Mammoth Cave became a national park.

Today, a total of 52,830 acres has been acquired;

known cave passages in the park total 350 miles. No one knows how long the cave system is, but it is the longest known system in the world. Most visitors come to Mammoth Cave for a cave tour; several different tours are offered daily ranging from walks on well-lit, paved paths to wild cave tours requiring crawling and squeezing through tight passages. Visitation in 2003 was 1,881,263.

The surface at Mammoth Cave is mostly second-growth forest with wetland and barrens intermixed. In addition to the park roads, there are seventy miles of trails through the park; thirty miles of the Green and Nolin Rivers flow through the park, providing opportunity for fishing as well as boating, canoeing, and camping along the river. Using one of these modes of transportation may bring a visitor into contact with several species of mammals.

The most numerous kind of mammal in the park is the rodent. The eastern gray squirrel is common everywhere. The eastern fox squirrel has been resident in the past, but few are seen today. The southern flying squirrel is extremely secretive. The eastern chipmunk resides in forested areas. In grassy areas, the groundhog, or woodchuck, may be encountered. Eastern woodrats nest near cave entrances.

The aquatic rodents include the beaver and the muskrat. Beaver were extirpated from Mammoth Cave long ago, but have been restocked in the park and are commonly seen in the Green River.

The eastern cottontail is the park's only lagomorph. The Virginia opossum is resident everywhere in the park. The raccoon is widespread. A reintroduction program has been started to bring the river otter back to the park.

The only large mammal at Mammoth Cave is the white-tailed deer. They were extirpated, but restocked in the 1940s; they are very abundant in the park today.

There are no large predators in the park, but several smaller ones are present. The bobcat is rare, nocturnal, and secretive. Two species of fox, red and gray, and the coyote inhabit Mammoth Cave. The coyote has been moving eastward and has now reached Mammoth Cave.

Several members of the weasel family live in Mammoth Cave National Park. Perhaps the most common is the eastern spotted skunk. They frequently forage in the campground. The less common striped skunk may also be seen around the campground.

Like many caves, Mammoth Cave is home to many bats. Twelve species reside in the cave. Bats are sometimes seen on cave tours, especially the eastern pipistrelle. Two of the bat species are on the federal Endangered Species List, the Indiana bat and the gray bat.

CONSERVATION CONCERNS

When Mammoth Cave was developed, the Historic Entrance became the focal point of cave activities. The campground, hotel, cabins, visitor center, and large parking lot are all near the Historic Entrance. Unfortunately, they are also built directly over the cave passages. Runoff from cars in the parking lot, including leakage of gasoline, antifreeze, and motor oil, may be finding its way into the cave. Efforts are being made to identify how much runoff is occurring and to suggest ways to prevent it.

Another water quality issue deals with the entire Mammoth Cave watershed. Most of the watershed lies outside park boundaries. Many potentially polluting land uses occur on private lands in the watershed including agriculture and forestry, oil and gas exploration, and commercial, industrial, and residential activities. Sewage and solid waste disposal occur. Non-point pollution caused by these activities may end up in the cave. Park staff must work with local authorities and private citizens to maintain the quality of park waters.

Air quality is routinely monitored in the park. Some of the problems are haze, ozone, and deposition. Haze is caused by particulates in the air, usually the result of soot or dust. The haze causes serious reduction in visibility and scenic quality. During summer, Mammoth Cave becomes one of America's haziest parks. It also ranks as one of the most ozone-polluted parks in the country. Ozone causes potential human health problems as well as damage to natural vegetation. Deposition involves various pollutants, including sulfates, nitrous oxides, and mercury, into both terrestrial and aquatic ecosystems.

The flora and fauna of Mammoth Cave are under stress. Among all species, seventy are listed as threatened or endangered on the federal or state list. Bat populations have been greatly reduced with two species now endangered. In addition, exotic flora has invaded the park. Pathogens have destroyed the American elm and American chestnut; the butternut is currently under attack. Exotic plants such as kudzu, honeysuckle, and tree-of-heaven are out-competing native plants. The loss of biodiversity will cause disruptions in the park's natural communities and the plants and animals that live within them.

The ruins of Mesa Verde's Cliff Palace include circular kivas, which were used for ceremonial purposes. The cliff dwellings of Mesa Verde were an important Native American habitat from 600 to 1300 A.D.

MESA VERDE

Land of the Cliff Dwellers

Cortez, Colorado

ACRES: 52,122

Most national parks were established for natural values, scenery, and wildlife. But Mesa Verde was preserved for its prehistoric Native-American culture. From 600 to 1300 A.D., people made a living in a semiarid environment by growing crops supplemented by hunting and gathering. Over those seven hundred years, food, clothing, and shelter evolved, culminating in the famous "cliff dwellings." Archeological studies at Mesa Verde are world renowned.

The Ancestral Pueblo people of Mesa Verde lived on a plateau now known as the "Green Table." It slopes downward from north to south with the southern edge cut by canyons. The northern part of the plateau rises to eighty-five hundred feet; the primary trees are gambel oak, Utah serviceberry, and mountain mahogany. In the south, pinyon pine and Utah juniper are most common. The canyon bottoms are characterized by yucca, prickly pear cactus, and sagebrush. Most of the canyons are dry; Ancestral people used springs and some flowing water for their supply.

Chapin and Wetherill Mesas are accessible by road; both contain significant cliff dwellings that are open to the public. Use of the park's 52,122 acres away from developed sites or designated trails is prohibited. This is one of the few national parks where substantial portions of the park's acreage are closed to the public.

The variation in elevation, about sixteen hundred feet, and availability of water create a diversity of habitats. There are more than sixty species of mammals in the park. Unfortunately, visitors may not be able to observe some of them due to closures.

Black bears exist in the park; campers frequently receive a warning about bears in Morefield Campground. Mountain lions also reside at Mesa Verde but are quite secretive. Bobcats are common but secretive.

Coyotes are widespread in the park, but are more heard than seen. On the mesa top, gray foxes are common but nocturnal. Red foxes are uncommon. The smaller carnivores include the western spotted skunk, striped skunk, ringtail, and badger.

Rodents are the most common mammals. Rock squirrels are common; the golden-mantled ground squirrel is limited to the oak woodland. There are two chipmunks in the park, the least chipmunk and the Colorado chipmunk. There are two tree squirrels as well, the red squirrel and Abert's squirrel. Three species of woodrat, or pack rat, reside in the park: the white-throated woodrat, the bushy-tailed woodrat, and the Mexican woodrat.

The Mancos River forms Mesa Verde's eastern boundary in the south half of the park. Occasionally beaver are seen in the Mancos; muskrats are more common. This is the only significant riparian habitat in the park. Porcupines live in Mesa Verde, but their populations are quite variable.

Rabbits and hares are frequently seen in the park. There are two species of cottontails, Nuttall's and desert. The black-tailed jackrabbit can be found in the arid area at the base of the plateau.

Among the larger mammals only the mule deer is common. The white-tailed deer and bighorn sheep were both extirpated from the park in the past. The bighorn was reintroduced in 1946 and is seen occasionally today. There is a small herd of elk in the region that occasionally wanders into Mesa Verde.

CONSERVATION CONCERNS

Mesa Verde received 434,813 visits in 2003. Since most of the use is highly directed and much of the park is closed totally, user impact is limited. There are problems, however. One of the serious problems is air pollution. Visibility from Far View Visitor Center is highly variable. Ship Rock, a noted landmark fifty miles south of Mesa Verde, is usually visible. But on days when haze is heavy, Ship Rock is nearly invisible. The cause is primarily pollutants from regional power plants. Monitoring air quality is continuing; acid deposition is not a major problem yet.

Mesa Verde was the site of several major forest fires in 2000 and 2002. Another fire burned in 1996. Most of the park was burned in these blazes. The combination of a hundred years of fire suppression and a prolonged drought, the park forests became highly susceptible to flame. On July 29, 2002, a lightning strike ignited a blaze in the pinyon-juniper forest on Chapin Mesa. About two thousand visitors were evacuated, and the park was closed for ten days. The possibility of further blazes at Mesa Verde and elsewhere due to fuel loadings from past suppression efforts remains high.

MOUNT RAINIER
Miles of Snow, Ice, and Lava

Ashford, Washington
ACRES: 235,625

Living in or visiting the Seattle area, it seems that Mount Rainier is always with you. The solitary snow-covered peak is located fifty miles southeast of the Seattle metropolitan area and is visible from many vantage points in the city. Mount Rainier, a dormant volcano that last erupted about 150 years ago, is the primary feature of Mount Rainier National Park. Steam vents are still found within the park. At 14,410 feet, Mount Rainier towers over the sea-level lowlands toward Seattle.

Mount Rainier National Park was created in 1899. It contains 235,625 acres, 97 percent of which is designated wilderness. About 58 percent of the park is forested, 23 percent is subalpine, and 19 percent is alpine, some of which is permanent ice and snow. The forests are primarily Douglas fir, western hemlock, and western red cedar. Red alder and black cottonwood are found along rivers. Timberline occurs between sixty-five and seventy-five hundred feet.

There is a tremendous elevation difference at Mount Rainier, approximately thirteen thousand feet, creating a wide variety of life zones and habitats. These habitats support a rich diversity of wildlife including approximately fifty-six species of mammals.

The black bear roams the park; their estimated population is 180. The other large predator in the park is the mountain lion. It lives in the forested areas. In 1999, there were thirty lion observations in the park. Smaller predators are seldom seen. These include the bobcat, red fox, and coyote.

There are several members of the weasel family (mustelids) in the park. The marten is perhaps the most common predator seen by visitors. Others include the long-tailed weasel, short-tailed weasel (ermine), mink, striped skunk, and western spotted skunk.

One of the most common mammals in the park is the American pika. The park's other lagomorph is the snowshoe hare. Snowshoe hares are found throughout the park up to alpine meadows.

The most commonly observed mammals at Mount Rainier are the rodents. The most common tree squirrel is Douglas's squirrel, which is found throughout the park's forested areas. The northern flying squirrel and the Cascade golden-mantled ground squirrel are resident. There are two chipmunks at Mount Rainier: Townsend's chipmunk and the yellow-pine chipmunk. The hoary marmot seeks out rocky areas.

The mammal most often noted by visitors is the black-tailed deer, a subspecies of the mule deer. They are found throughout the park up to timberline. During winter, they migrate to the lower elevations.

In the eastern part of the park, elk are sometimes seen. They were extirpated from the park in the late nineteenth century. They are now reappearing; it is believed that they are descendants of elk introduced elsewhere in the Cascades that are now expanding their range.

One of the highlights of a visit to Mount Rainier is seeing a mountain goat. It is possible to see them almost anywhere there are cliffs or rocky slopes. They may be active day or night. Goats are social animals and may be seen in groups of up to forty individuals. A park survey estimated the goat population at four hundred.

CONSERVATION CONCERNS

Mount Rainier is located close to the Seattle metropolitan area, one of the fastest growing areas in the country. The state of Washington is projected to double in population by the middle of this century. In 2003, Mount Rainier registered 1,312,415 visits; a significant portion of it from the nearby urban areas.

Use of Mount Rainier will likely grow substantially in the next few years. A major concern for management is maintaining the quality of the park's natural resources as well as the visitors' experience. Decisions for future management will be critical.

Proximity to a large urban area brings with it the threat of air pollution. As noted above, growth continues in the Seattle-Tacoma region, bringing with it increases in pollutants. Pollutants also come from Vancouver to the north and Portland to the south. Long-distance movement of pollutants from Asia and Europe is also occurring. Sources include industry and motor vehicles, which contribute ozone, sulfates, and particulates.

The problems that result include reduced visibility, direct impacts on ecosystems especially vegetation, and potential human health problems. Air quality is being monitored, specifically measuring impacts on visibility, aquatic ecosystems, and vegetation. Though the Clean Air Act considers Mount Rainier a Class I area, air quality solutions will depend on convincing polluters of the necessity of clean air.

Mount Rainier staff has been monitoring aquatic ecosystems since 1988. There are concerns about the biological and chemical characteristics of the park's waters. Some of the issues being assessed are acid deposition, the effects of stocking non-native fish, and the effects of recreation visitation.

Mount Rainier has been the site of considerable development in the past. Camping and picnicking facilities were built throughout the park including fragile alpine meadows. A golf course and stables were built there as well. A recent management thrust in parks has been to rehabilitate developed sites to their natural condition where possible. Such a project is underway at Mount Rainier. Canon USA, Inc., is providing partial funding for these efforts.

Located only fifty miles from Seattle, Mount Rainier is one of the earliest national parks. An extensive trail system provides even the most casual of hikers a variety of views of the snow-covered summit.

NORTH CASCADES
Land of the 700 Glaciers

Marblemount, Washington
ACRES: 684,302

The area now known as North Cascades National Park was first proposed as a national park in 1892. Local residents were concerned about uncontrolled sport hunting for grizzly bears and mountain goats. Most of these local residents

were miners and represented the largest population the area ever had. However, with the discovery of gold in the Klondike, most of the local miners left and the issue died.

In 1891 and 1897, Congress passed the Forest Reserve Acts, which authorized the president to declare federal forest lands as "forest reserves," closing them to homesteading and retaining them in federal ownership. In 1897 President Grover Cleveland declared the North Cascades area the Washington Forest Reserve. Subsequently, the area was divided into the Mount Baker, Okanogan, and Wenatchee National Forests. Forest Service ownership continued until 1968 when the North Cascades National Park Complex was established under National Park Service management.

The North Cascades National Park Complex includes three areas managed by a single staff. North Cascades National Park contains 684,302 acres and is essentially roadless. The park is divided into two parts, separated by Ross Lake National Recreation Area (117,574 acres). To the south is the Lake Chelan National Recreation Area, which contains 61,890 acres. Over 93 percent of the complex was designated as wilderness by Congress in 1988. The North Cascades Ecosystem includes considerable wilderness adjacent to the park to the north in British Columbia as well.

There are two reasons the three-part complex was created. First, most manmade developments were placed in the national recreation areas, while the rugged wilderness with only trail access was placed in the national park. Both national recreation areas contain impounded lakes. Ross Lake National Recreation Area contains three impoundments including twelve thousand-acre Ross Lake. Lake Chelan is Washington's largest natural lake and the nation's third deepest. However, in 1927, its outlet was dammed to create hydroelectric power. Second, hunting is allowed in both national recreation areas but is prohibited in the national park.

North Cascades is wild mountain country. More than three hundred glaciers cap the high country along with countless snowfields. The west side of the Cascades gets more than five hundred inches of snow annually; east of the mountains, snowfall is only about 110 inches. Vegetation reflects climatic differences. West of the Cascades, the forests contain Pacific silver fir, western hemlock, western red cedar, and Douglas fir. On the drier side, the most important trees are lodgepole and ponderosa pine. In the high country, alpine meadows abound.

Mammals in North Cascades reflect the wilderness nature of the park. Several threatened and endangered mammals are included. Seventy-five mammal species live in the park.

The grizzly bear is one of the park's threatened species. An estimated thirty to fifty grizzlies live in the entire North Cascades Ecosystem. Black bears also live throughout the park. The mountain lion prefers habitats of dense forest. The population of the bobcat is stable, and its distribution is widespread.

There are three canid predators in North Cascades including the endangered gray wolf. Wolves have been seen from time to time in the complex and on adjacent lands in Canada. The red fox is typically found in the forests; the coyote is commonly heard.

The weasel family is well represented in the park, including the short-tailed weasel (ermine), long-tailed weasel, and marten. Mink and river otter live in the lower elevations. There are two skunk residents of the park, the western spotted skunk and the striped skunk.

The American pika inhabits the entire park wherever there are rocky outcrops. The other lagomorph in the park is the snowshoe hare.

Among the tree squirrels, the most commonly seen is Douglas's squirrel. The red squirrel and the western gray squirrel, a state threatened species, are residents. Ground-dwelling squirrels are also common including the Cascades golden-mantled ground squirrel. There are also two chipmunks, Townsend's chipmunk and the yellow-pine chipmunk.

There are several ungulates in the park. The mule deer and its black-tailed deer subspecies both inhabit the park. Moose are inhabitants of lower elevations. Elk are rare in the park, having moved in from adjacent areas.

Mountain goats are sometimes seen by visitors to the high country. They prefer windy areas such as cliffs or snowfields.

CONSERVATION CONCERNS

Remote rugged wilderness doesn't always mean pristine wilderness. Prevailing winds at North Cascades come from the southwest, directly from urban Puget Sound. Pollutants include sulfates, nitrates, particulates, and ozone. Some pollutants are deposited on glaciers. Glacial runoff carries pollutants into aquatic ecosystems. Some pollutants are deposited directly into rivers and lakes. Deposition can cause serious problems for aquatic organisms

(Pages 70–71) Ruby Beach at dusk, Olympic National Park.

Packing with horses is a popular pastime in the backcountry of North Cascades National Park. North Cascades is actually a park complex, made up of almost seven hundred thousand acres of virtually roadless wilderness.

by reducing pH levels. Plankton, the base of the food chain, is highly sensitive to changes in acid. The effects are felt throughout the ecosystem. Some fish, including salmonids, are sensitive to increased acid, causing difficulties in spawning.

Other sources of pollution come from the Fraser River Valley in British Columbia. Farms, industry, and vehicles help create high levels of ozone, which moves into the park, especially in summer. Ozone is potentially a serious human health hazard. It also damages the foliage of plants and reduces productivity and vigor.

Air and water resources have not been seriously degraded to date. Monitoring of air and water quality is continuing in conjunction with the U.S. Geologic Survey. Pollutants being monitored are ozone, sulfur dioxide, nitrogen oxides, organic carbon, and fine particulates. Acid precipitation and visibility are measured by photography.

There are 258 exotic plant species documented in the park. Diffuse knapweed, which has the ability to spread considerably, causes the biggest problem. It crowds out native plants in shrub and grassland ecosystems; wildlife usually will not eat it. Control efforts involve physical removal by pulling, biological control, and, as a last resort, spraying.

OLYMPIC

The Majestic Beauty of Washington State

Port Angeles, Washington
ACRES: 922,651

You might not think that the location in the lower forty-eight states that gets the most precipitation would be a major tourist attraction. But if there are rugged mountains, a spectacular seacoast, and rain forests with trees over three hundred feet tall, perhaps the rain gets overlooked. Ignoring the forty feet of snow in the high elevations of the park, however, may be more difficult. Nonetheless, in 2003, more than three million people came to Olympic National Park.

Located on the Olympic Peninsula, Olympic National Park contains 922,651 acres in two parts. The major part of the park surrounds Mount Olympus (7,965 feet) and its glaciers. Lower elevations contain rain forests of western red cedar, western hemlock, Sitka spruce, and Douglas fir. Timberline in the park is usually about five thousand feet. Above that are alpine meadows that are frequently covered with flowers.

The second part of the park is the Pacific Ocean seacoast, stretching from Ozette to Kalaloch. Facing west, the wind-buffeted coast is forest covered. Marine mammals are sometimes seen from vantage points along the beach.

Olympic National Park has an interesting political history. The original reason for protection was to preserve the declining population of elk in the Olympics. In 1909 Theodore Roosevelt designated Mount Olympus National Monument in the Olympic National Forest. From 1909 to 1933, the monument was managed by the U.S. Forest Service. In 1933, national monuments were transferred to the National Park Service. In 1937, Franklin Roosevelt visited the monument and supported the idea of enlarging the monument and giving it national park status in 1938.

Today, Olympic preserves habitats for other wildlife, including fifty-four species of mammals. The Roosevelt elk remains in the park. It is estimated that seven thousand elk live on the Olympic Peninsula. There is a small herd resident in the Hoh Rain Forest; in winter, they are especially visible. The Columbia black-tailed deer is very common, living everywhere from the beach to the high country.

The largest predator in the park is the black bear. Both the mountain lion and bobcat are common. The heaviest population density of mountain lions in the United States is in the Olympic Peninsula. The bobcat is nocturnal and extremely secretive. The coyote is frequently heard. Raccoons are common in low elevation areas and along the coast.

There are numerous members of the weasel family in the park. The common Olympic short-tailed weasel, a subspecies, is endemic to the park. The long-tailed weasel, mink, river otter, and marten are resident.

The snowshoe hare is found anywhere below timberline, including along the beaches. Oddly, the park's hares do not change color in winter as most do.

The number of rodent species at Olympic is less than in other large parks. The Olympic marmot is

The woods of Olympic National Park contain America's temperate rain forests, but the park is also famous for its meadows, glaciers, and shoreline. Driftwood often washes up from the adjacent Pacific Ocean.

endemic. Townsend's chipmunk is common and can be found at overlooks and in campgrounds. The park's tree squirrels are Douglas's squirrel, or chickaree, and the northern flying squirrel.

Pacific Ocean waters near shore to Olympic National Park contain numerous marine mammals. The whales along the coast are migrants; the most common of them is the gray whale, which is seen passing close to shore in spring and fall. A little farther offshore are two other migrant whales, the Minke and the humpback. Two smaller cetaceans, the Dall's porpoise and the Pacific white-sided dolphin, can be seen in lagoons and river mouths.

The harbor seal is the most abundant of the marine mammals. Northern sea lions (Steller sea lions) and California sea lions migrate through the park's waters. The sea otter was overhunted and extirpated many years ago off the coast of Olympic. In 1969, they were reintroduced and their numbers are today increasing.

CONSERVATION CONCERNS

The National Park Service continues to follow a policy of rehabilitation and restoration of park ecosystems. Such policies attempt to create a vignette of pre-Columbian America, meaning some species would need to be reintroduced if possible, while others might have to be removed. Both cases are true at Olympic.

The mountain goat is an introduced species in the park. A dozen of them were brought to the park more than a hundred years ago from Alaska and Canada. The population has now grown to several hundred, greatly exceeding the carrying capacity of the park. Backcountry impacts are most evident. Serious consideration is now being given to removing the goats from the park, which would certainly be controversial.

The gray wolf was once part of the fauna of Olympic, but the last recorded sighting was in 1922. As in Yellowstone, a proposal is in the works to reintroduce the wolf.

Discussion has also been under way concerning the removal of dams from the park. Several dams were installed before Olympic became a national park. Removing them would restore naturalness in aquatic ecosystems and would have a substantial impact on terrestrial ecosystems as well.

The mountain portion of Olympic National Park is mostly surrounded by the Olympic National Forest. In a positive sense, unwanted developments, such as residential and commercial sites on the park's boundaries, are restricted. Conversely, some of the extractive activities that are allowed on national forests may cause impacts on visitation, especially logging.

Long-distance movement of pollutants is a potential problem for Olympic. Deposition of pollutants in water, on land, or remaining in the air must be monitored; their effects must be identified as well. To date, these problems have not become serious.

PETRIFIED FOREST
Where Science and Scenery Come Together

Petrified Forest, Arizona
ACRES: 93,533

A few decades ago, historic Route 66 crossed Arizona on its journey from Chicago to Los Angeles. Many travelers tried to avoid the heat of the desert by driving their non-air-conditioned autos through it at night. But they missed some interesting scenery by doing so, including the Petrified Forest. Today, however, most tourists travel in cool comfort and are willing to stop and see nature's handiwork.

Petrified Forest National Park contains 93,533 acres of desert including the colorful Painted Desert. What the more than half million visitors are seeking, however, is petrified wood. About 225 million years ago, this area was well watered and heavily forested, inhabited by an assortment of strange amphibians and reptiles. Most of the petrified trees were from *Araucarioxylon arizonicum*, a species similar to modern-day Norfolk Island pine.

Eventually, through disturbance and climate change, the ancient forest was buried and petrified, the organic bodies of the individual trees turned to stone. Today, their remains can be seen in the Crystal Forest, Jasper Forest, and Rainbow Forest, each name reflecting the striking colors in the now unburied petrified trees.

Native Americans inhabited the area for ten thousand years and left some evidence for archeologists. Europeans arrived in the seventeenth century to find high desert and short grass prairie, much of it covered by sagebrush. Few trees blocked the landscape. Annual precipitation is less than nine inches, half of it coming in strong summer thunderstorms.

In 1882, the Santa Fe Railroad was built across Arizona, opening the region to tourism. Huge vol-

Pronghorns are found in grassy areas; groups of up to fifty winter in the park. The mule deer is seen infrequently.

CONSERVATION CONCERNS

The park's biggest concern is theft of petrified wood. It is so tempting for visitors to pick up small pieces. Despite interpretive efforts to explain the harm theft causes and substantial law enforcement efforts, it is estimated that twelve tons of wood are stolen annually. Even with a minimum fine of $275 for petrified wood theft, people continue to risk being caught.

Pollution problems are not major issues in the park to date. There are ten surface drainages in the park, many of which are dry for portions of the year. No major water quality concerns have been found. There are power plants in the region, but prevailing winds usually carry pollutants elsewhere. Air quality in the park meets current ambient quality standards. Future developments could change the situation, however.

As in most other parks, the presence of exotic plants is a problem at Petrified Forest. Control is difficult and costly once an exotic population is established. Vigilance in identifying new infestations is key.

Petrified Forest has seen the outbreaks of mammal-related diseases in the past. These include rabies, plague, and hantavirus, all of which are potentially fatal in humans. Visitors need to avoid dead or sick animals and to prevent pets roaming free. It is best to not approach any wildlife, especially any animal that is acting out of the ordinary.

Erosion has produced extraordinary colors and shapes in Petrified Forest National Park. After each rain, new specimens of petrified wood are exposed.

umes of petrified wood were collected and sold for souvenirs. Concerned for how long the petrified wood would last, President Theodore Roosevelt established Petrified Forest National Monument in 1906 under the authority of the Antiquities Act. Petrified Forest achieved national park status on December 9, 1962.

Today, the major effort at Petrified Forest is to interpret these unique resources to the public while protecting it for future generations. A twenty-eight-mile paved road connects many of the park's attractions including interpretive trails that provide access to some of the most extraordinary deposits of petrified wood.

Even though Petrified Forest focuses on geology, it has an interesting variety of mammals. It is estimated that forty-five mammal species inhabit the park. The most numerous species are the rodents and bats, though many of the latter are of questionable residency. Conversely, no large predators are resident in the park. Most desert mammals are nocturnal to avoid the heat of the day as well as predators.

The coyote is active at dusk and dawn. The ringtail and raccoon are regular park residents but are nocturnal and seldom seen. There are three members of the weasel family (mustelids) in the park. All are year-round residents: badger, western spotted skunk, and striped skunk.

There are two species of lagomorphs in the park, both of which are out at dusk and dawn. The desert cottontail and the black-tailed jackrabbit may be seen along roads early and late in the day.

Rodents are well represented in the park. Probably the most observed is the white-tailed antelope squirrel. Others include the spotted ground squirrel and several species of mice and rats. Ord's kangaroo rat never drinks; it gets all the moisture it needs from seeds it consumes. A colony of Gunnison's prairie dogs inhabits 160 acres of the park.

REDWOOD
Trees that Touch the Sky

Crescent City, California

ACRES: 112,513

The battle to save the redwoods was a long process resulting in the creation of Redwood National Park. Saving the redwood is certainly worthwhile; a 367.8-foot redwood is the world's tallest tree. Redwood stands are awe inspiring.

As forests of the Midwest began to run out, the logging industry looked to the northwestern part of the United States as the next source of supply. In

Sunlight filters through a grove of the "world's tallest trees" in Redwood National Park. The redwoods in the park often reach three hundred feet in height, with the tallest being 367.8 feet.

1850, two million acres of old-growth redwood existed in a narrow band along the coast of northern California. Logging began in the 1850s, and in 1878 federal legislation provided for the sale of federal timberland to private interests for pennies per tree. Old-growth redwood was included. By the turn of the twentieth century, more than half the redwoods were gone.

The public soon realized conservation efforts were necessary. The first success was in 1902 when Big Basin Redwoods State Park was created. At the federal level, Theodore Roosevelt was instrumental in the establishment of Muir Woods National Monument near San Francisco in 1906. Muir Woods contains one of the southernmost stands of redwood.

The redwood harvest continued apace. In 1918 the Save-the-Redwoods League was established to expand protected redwoods. The League supported the creation of a California state park system and a process for funding. In the 1920s, three state parks were established containing some of the finest redwood trees remaining—Prairie Creek, Jedediah Smith, and Del Norte Coast State Parks. The three total twenty-seven thousand acres, twenty-four thousand of which were old-growth redwood.

Was the acreage of protected redwoods sufficient? Logging continued outside the state parks and by the 1960s, 90 percent of the redwoods had been logged. The environmental movement of the 1960s gave impetus to preserving more redwood acreage. In 1968, Redwood National Park was created, saving fifty-eight thousand additional acres of redwood. The new park included thirty-seven miles of Pacific Ocean coastline. The three state parks noted above were to be included in the national park; the state of California refused.

Ten years later, the park was enlarged again by forty-eight thousand acres, thirty-nine thousand acres of which had already been logged. Finally, in 1994 the National Park Service and the California Department of Parks and Recreation agreed to manage the national park acreage and the three state parks cooperatively. These lands contain 45 percent of the remaining old-growth redwoods.

Several species of mammals inhabit the park including marine mammals along the coast. There are two species of ungulates in the park; Roosevelt elk and black-tailed deer (a subspecies of mule deer) are common. The redwood forests are interspersed with grassy prairies, which provide quality habitats.

Another resident of the grassy prairies is the black bear. It is common even along the beach. The other large predator is the mountain lion. Bobcats are common. Gray foxes and coyotes reside throughout the park. The western spotted skunk and striped skunk are common. The long-tailed weasel, short-tailed weasel (ermine), marten, river otter, and ringtail are park residents.

Ground squirrels include the California ground squirrel and the golden-mantled ground squirrel. The park's tree squirrels are the western gray squirrel and Douglas's squirrel, or chickaree. Three chipmunks inhabit the park: Allen's, Siskiyou, and Sonoma. Redwood is the only national park that hosts the latter two species. Porcupines are very common in the park's forests.

Among the lagomorphs, the brush rabbit is quite common. The black-tailed jackrabbit inhabits grassy areas.

Numerous marine mammals are visible from the coast. The only one resident in the park is the harbor seal, which is common along the park's ocean frontage. Several other pinnipeds are frequently seen. These include the northern (Steller) sea lion, California sea lion, and northern elephant seal.

Gray whales migrate past the park in spring and fall and can be seen regularly from park overlooks. Other whales that are sometimes seen include sperm, sei, Minke, fin, humpback, goosebeaked, killer (orca), and false killer whales. Smaller cetaceans include the short-beaked saddleback dolphin, Dall's and harbor porpoise, and the Pacific white-sided dolphin.

CONSERVATION CONCERNS

Redwood National Park monitors air and water quality. There are no large municipalities in the area, but airborne pollutants can cause serious problems. To date, no significant problems have appeared. The park attracted 406,058 visitors in 2003; physical impacts from visitation are currently under control.

The major park issue deals with the rehabilitation of logged-over lands acquired in 1978. These acres had to be stabilized to prevent erosion and sedimentation. Former logging roads are being returned to their original contours as are skidding trails. Prairies are being returned to their original contours and original vegetative cover. Deer and elk are expected to return to these areas. Finally, cutover areas will eventually be replanted with native species.

Upstream from the park along Redwood Creek is the Park Protection Zone. The zone contains thirty thousand acres of private land where National

Park Service resource managers will help prevent the negative effects of logging and other land uses downstream. Giving the Park Service input in private land-use decisions is a novel approach.

Park managers are also returning fire to park ecosystems to improve landscape health. For prairies and oak woodlands, fire is being used to control exotic vegetation, remove intruding conifers, and restore native plant diversity. In old-growth forests, fire is being used to return disturbance and to reduce fire-intolerant species. The ultimate goal is to restore as much naturalness to the landscape as possible. Fire is an important tool in achieving that goal.

ROCKY MOUNTAIN
The Spectacular Peaks of Colorado

Estes Park, Colorado

ACRES: 265,828

Rocky Mountain National Park contains extraordinary mountains. Some people think it is the site of the tallest mountain outside Alaska—no, that's in California. Some think it is the site of the tallest mountain in Colorado—no, that's Mount Eldon, west and south of Rocky Mountain. The park does have plenty to offer: sixty peaks above twelve thousand feet and the highest continuous paved auto road in the United States, Trail Ridge Road, which reaches 12,183 feet in elevation.

The region that would ultimately become Rocky Mountain National Park was purchased from France as part of the Louisiana Purchase of 1803. Miners, homesteaders, and tourists helped develop the area in the nineteenth century. Led by naturalist Enos Mills, efforts soon began to create a national park; success was achieved in 1915. Today, Rocky Mountain National Park contains 265,828 acres of some of America's best scenery.

The park contains three major ecosystems. The montane ecosystem is found below ninety-five hundred feet. On south-facing slopes, ponderosa pine is the prominent tree. Grasses and shrubs are intermixed. On north-facing slopes, Douglas fir and lodgepole pine are most common. Higher, up to eleven thousand feet, is the subalpine ecosystem. The ma-

jor tree species here are subalpine fir, Engelmann spruce, and limber pine. At timberline, the forest becomes *krummholz* ("crooked wood"), trees that grow only as tall as protecting rocks allow; high winds prohibit any further height growth. Finally, above timberline is alpine tundra where the only tree is dwarf willow. Approximately one-third of the park is above timberline.

Elevations in the park range from seventy-seven hundred feet up to 14,225 feet at the summit of Longs Peak. There are more than 350 miles of trail in the park including the Longs Peak Trail to the summit. There is ample opportunity to see the park's sixty-six mammal species by foot or park road.

In historic times, the grizzly bear inhabited the park, but it has been extirpated, leaving the title of largest predator to the black bear. The other large predator in the park is the mountain lion. Among the canids, the gray wolf has been extirpated as well. The coyote is the most common canid. The mustelids (weasel family) at Rocky Mountain include the badger, pine marten, long-tailed weasel, river otter, and striped skunk. Visitors are not likely to see any of them.

Among the lagomorphs, the American pika lives in the high country. Snowshoe hares, white-tailed jackrabbits, and Nuttall's cottontail are residents.

Rodents abound in Rocky Mountain. There are two species of chipmunks, the least chipmunk and Uinta chipmunk. Ground squirrels include the golden-mantled ground squirrel, Richardson's ground squirrel, and the yellow-bellied marmot. Rocky Mountain's forests support the red squirrel, or chickaree, the handsome Abert's squirrel, and the porcupine.

Perhaps the highlight of wildlife viewing at Rocky Mountain is the ungulates. Elk are common in the park. This was not always so. With settlement in the Estes Valley, elk were almost extirpated. In 1913 and 1914, forty-nine elk were moved from Yellowstone to Rocky Mountain National Park. About the same time, predator control programs removed many of the carnivores that prey on elk. Today's elk population varies seasonally in the park. About thirty-two hundred elk live in the park in summer; as few as a thousand remain during winter.

Bighorn sheep were plentiful in the region until hunters and settlers came on the scene. Both sport and market hunters reduced the population as ranchers took over bighorn habitat for their domes-

Winter comes early and stays long in the high country of Rocky Mountain National Park. Snow remains into the summer months. Visitors may enter the park in winter, but the famous Trail Ridge Road is closed to auto traffic once the snows arrive.

tic sheep. The domestic animals became a source of disease for the bighorns. By the 1950s, only 150 bighorns remained in the park. With the decline in hunting and disease, bighorn populations rebounded. The population today may be as many as six hundred.

One concern for bighorns is the potential presence of mountain goats in the park, which may carry diseases that could harm bighorns. There is an introduced population of mountain goats in the Mount Evans area, south and west of Rocky Mountain, that occasionally wander into the park. They are typically captured and returned to Mount Evans but at times, when capture isn't possible, park policy permits shooting.

Moose apparently were only transients in the park historically. In 1978 and 1979 the Colorado Division of Wildlife introduced twenty-four moose into the park west of the Divide. They have prospered and spread since then and now are commonly seen by visitors.

Mule deer are also present in all park ecosystems. They typically spend the summer in meadows at higher elevations while in winter they move lower.

CONSERVATION CONCERNS

Rocky Mountain National Park is one of America's most popular with an annual visitation of three million people. Being close to Denver and the high-growth area along the Front Range suggests visitation will increase. Unfortunately, resource issues go beyond high visitation. Air quality has been monitored since 1990. Ozone has increased since monitoring began. Ozone can cause human health problems especially among people with respiratory problems. It also causes damage to vegetation.

Another increasing problem is the presence of nitrous oxides, mostly caused by motor vehicles from along the Front Range as well as industrial and agricultural sources. Nitrous oxides are not only a factor in the creation of ozone but also lead to nitrate deposition. Nitrates in the park's precipitation have increased since 1990, causing changes in soil and water chemistry and, in turn, changes in the food chain.

On the positive side, sulfates have decreased since 1990. Sulfates come from the burning of coal; power plant technology has improved, thus reducing sulfates. Sulfates are a major cause of acid deposition and a significant factor in the reduction in visibility. Problems related to deposition and visibility have improved over the last decade.

Water quality in the park is high except for nitrogen deposition noted above. The major concern, however, is that as natural buffering is used up, deposition issues will become more serious.

Finally, there is a new concern for park biologists, chronic wasting disease (CWD), a neurological disease in deer and elk. CWD has been found in deer in northeastern Colorado at rates from 1 to 11 percent. Deer are being tested within the park to determine if or when CWD will become present in park ungulates. An outbreak of CWD within the park's deer or elk is horrible to contemplate.

SAGUARO
Land of the Cactus

Tucson, Arizona

ACRES: 91,440

Most Americans are aware of the cactus "with arms." They see them in movies and cartoons but may not know this cactus' real name. The "armed cactus" is the saguaro, namesake to Saguaro National Park. Saguaro preserves a portion of the Sonoran Desert, the only American desert where saguaros grow. The park is divided into two parts, one on either side of Tucson, Arizona. The Tucson Mountain Unit, west of Tucson, protects the area around the highest peak in the Tucson Mountains, Wasson Peak (4,687 feet), as well as the surrounding desert. The eastern portion, the Rincon Mountain Unit, includes Sonoran Desert lowlands up to the summit of Mica Mountain (8,666 feet).

Saguaros are found only up to an elevation of about four thousand feet. Freezing temperatures in winter are one of the saguaro's limiting factors; frost is seldom seen below four thousand feet. Most of the Tucson Mountain Unit is desert scrub including saguaros; the higher areas are in the grassland transition zone. The Rincon Mountain Unit contains significant desert scrub and grassland too, but most of it is forest. Ecosystems include oak-pine woodland, ponderosa pine, and a small amount of Douglas fir forest, especially around Mica Mountain.

The Tucson Mountain Unit is limited to day use. Several roads and trails allow access to the unit. Most of the Rincon Mountain Unit is inaccessible to cars. There is a loop drive, the Cactus Forest Drive, giving visitors a close-up look at the saguaro forest. Getting to the high country requires a walk that gains several thousand feet in elevation. Total park acreage is 91,440.

tractive for residential development. It has already begun at Saguaro; it will happen elsewhere.

The presence of the city also means traffic congestion and pollution. Emissions from motor vehicles decrease air quality at times in the Tucson basin. There are other sources of pollution in the region. Copper smelters contribute to air degradation, especially sulfates.

More than six hundred acres of the park have been infested with exotic flora. Recent modeling has shown that if untreated, invaded land will increase to twelve thousand acres. A program of control is underway that includes mechanical, cultural, biological, and chemical controls.

Ranks of cacti seem to march up mountain flanks in Arizona's Saguaro National Park. These are the saguaro cactus made famous in movies and cartoons. Only saguaros over seventy-five years old have the familiar "arms."

The park supports a variety of habitats and a variety of mammals. Rodents may be the most commonly seen mammals. Harris's ground squirrel is abundant in the desert and foothills. Lower, the round-tailed ground squirrel is abundant. In the higher forested areas of the Rincon Mountain Unit, two tree squirrels are resident, the Arizona gray squirrel and the tassel-eared Abert's squirrel. Two species of woodrat inhabit the park, the white-throated woodrat and the Mexican woodrat.

Among the lagomorphs, the desert cottontail is abundant. There are two jackrabbits in the park, the black-tailed jackrabbit and the antelope jackrabbit.

There are three species of ungulates in Saguaro, the mule deer, the white-tailed deer, and the collared peccary, or javelina. This last ungulate travels in groups of up to fifty individuals. They can be seen moving through the park's desert and foothill areas.

The most common predator in the park is the coyote. The gray fox lives mainly in the desert and foothills; the rare kit fox is a desert dweller. There are two felid species at Saguaro, the bobcat and the mountain lion; they range in all habitats but are very secretive. A few black bears reside in the park, mostly in the higher elevations.

The weasel family at Saguaro consists of four species of skunks and the badger. They include the striped skunk, western spotted skunk, hooded skunk, and western hognosed skunk.

CONSERVATION CONCERNS

When Saguaro was established as a national monument in 1933 (national park status was declared in 1994) the two units were beyond the edges of Tucson. As the city grew, eventually it butted up to the park. Today, the west edge of the Rincon Mountain Unit abuts residential areas. Where private property exists adjacent to a national park, it becomes very at-

SEQUOIA–KINGS CANYON
Home to the Oldest Living Things

Three Rivers, California
ACRES: 865,952

The world's first national park, Yellowstone, was established in 1872. The United States took its time creating another. In the meantime Australia established Royal National Park in New South Wales, and Canada created Banff National Park in Alberta. But in 1890, three new parks were established in the Sierra Nevada in California. America's second national park was Sequoia National Park, followed within a week by General Grant National Park and Yosemite National Park. Later, General Grant National Park was combined with the new Kings Canyon National Park (460,136 acres), taking the name of the latter in 1940. Today, since they share a long boundary, Sequoia and Kings Canyon National Park are being managed as one unit totaling 865,952 acres.

Both Sequoia and General Grant National Parks were created out of the concern that some of the world's largest trees, the giant sequoias, would be logged. A sawmill had been built just nine miles from the Giant Forest, forcing Congress to act. Sequoia National Park protected the Giant Forest, which includes the General Sherman and Chief Sequoyah trees. The General Sherman Tree is over two thousand years old. It is 275 feet high and 103 feet in circumference and is considered by most people to be the world's most voluminous living thing. General Grant National Park was the site of the Grant Grove containing the General Grant Tree, which is 267 feet high, 108 feet in circumference, and second only to General Sherman in volume.

Sequoia–Kings Canyon National Park is more than just big trees. On the eastern boundary of Sequoia is the tallest peak in the country outside Alaska, 14,491-foot Mount Whitney. Kings Canyon National Park contains the spectacular canyons of the Kings River. There is an elevation difference in Sequoia–Kings Canyon from thirteen hundred feet up to the summit of Mount Whitney.

Mammals in the park are typical of those in the Sierra; a total of seventy species is thought to exist in the parks. Black bears are common; mountain lions are rarely encountered. The coyote is common in both Sequoia and Kings Canyon. Both the gray fox and red fox are residents. The weasel family in Sequoia–Kings Canyon includes the short-tailed weasel (ermine), marten, long-tailed weasel, western spotted skunk, striped skunk, and fisher. Raccoons are common; ringtails are found around rocks and caves.

The most frequently observed mammals are the diurnal rodents. The California ground squirrel, the golden-mantled ground squirrel, and Belding's ground squirrel are present. There are four species of chipmunks—the lodgepole chipmunk, the alpine chipmunk, Merriam's chipmunk, and the Colorado chipmunk.

The largest member of the squirrel family here is the yellow-bellied marmot. Tree squirrels inhabit the forested portions of the parks and include the western gray squirrel and Douglas's squirrel. Porcupines also live in the forested areas. Three species of woodrat are found in the parks: the dusky-footed woodrat, the bushy-tailed woodrat, and the desert woodrat. Mountain beavers reside in the montane riparian areas. Beaver and muskrat are resident but not native; both were introduced.

The American pika is the most often seen lagomorph. Other lagomorphs include the desert cottontail, brush rabbit, black-tailed jackrabbit, and white-tailed jackrabbit.

Sequoia–Kings Canyon has few ungulates. The mule deer is common; bighorn sheep live in alpine areas and are almost never seen by visitors.

CONSERVATION CONCERNS

Numerous conservation issues are relevant to Sequoia–Kings Canyon, including air pollution,

exotic species, water pollution, loss of natural fire regime, and habitat fragmentation. Perhaps the most serious air pollution issue is ozone. Ozone is produced when pollutants from industry and motor vehicles are heated by sunlight. Ozone is dangerous to human health, especially to individuals with respiratory problems. In 2001 Sequoia–Kings Canyon recorded more days with unhealthy levels of ozone than any other national park—sixty-one days.

Ozone is also harmful to vegetation. Jeffrey and ponderosa pines are susceptible to ozone damage. Ninety percent of the Jeffrey pines in the Giant Forest area show ozone damage. Reduced photosynthesis and slower growth rates are the result. Sequoia trees are more resistant, though some scientists think ozone affects sequoia seedling survival.

Acid deposition is not a major problem as yet, but high elevation lakes are known to be susceptible to acidification. As human population of the nearby San Joaquin Valley grows, acid deposition will likely become an issue. Nitrogen deposition is increasing, probably caused by vehicles, industry, and agriculture. It is also known that more nitrogen is being retained in the parks' vegetation. The long-term results of this are undetermined.

Finally, agricultural chemicals such as organophosphates and PCBs become suspended in air as particulates and are being deposited in the park. The cause-and-effect relationship has not yet been proven, but some scientists believe this deposition

Once separate, Sequoia and Kings Canyon National Park are managed together and share a common boundary. The groves of Giant Sequoia include the nation's "most voluminous tree," the General Sherman Tree, which has a circumference of 103 feet.

is causing harm to the parks' wildlife. Peregrine falcons on Moro Rock in Sequoia National Park have never produced offspring; the reason has not yet been identified. Pesticide drift is thought to be the cause of the extirpation of two of Sequoia National Park's frog species—the foothill frog and the mountain yellow-legged frog.

Reduction in visibility is becoming a greater problem. The farthest object visible in the park is 105 miles away in the Coastal Range. According to the Environmental Protection Agency (EPA), the average natural visibility in the western United States should be 120 to 180 miles. Yet according to the National Parks Conservation Association, at Sequoia-Kings Canyon annual visibility averages just thirty-nine miles. The culprit is particulates, typically from soot from diesel engines and power plants, and dust.

Exotic flora and fauna are also problems. Some exotic plants take over local areas, creating monocultures and destroying diversity. Sequoia–Kings Canyon is plagued by exotic wildlife. As noted above, the exotic beaver is resident in the park. Beavers' engineering feats allow them to alter environments; these activities are monitored to minimize such alteration. The park is resident to several feral mammals. Feral cats destroy native wildlife; the rooting of feral pigs ruins soil. Domestic cattle have grazed on native vegetation to the point where, in 2001, parts of the park boundaries were fenced to keep them out.

The High Sierra wilderness in Sequoia–Kings Canyon contains thirty-two hundred lakes and ponds and about twenty-six hundred miles of rivers and streams. The park is an important source of water for downstream users. Air pollution is the biggest worry regarding water quality. Deposition of acids and pesticides in park waters is occurring. These park waters are poorly buffered, though acidification of streams and lakes is not serious to date.

Sequoia–Kings Canyon attracts more than 1.5 million visitors annually. Where backcountry use proliferates, water quality can become an issue. Downstream from popular campsites, waters may contain human fecal material, soaps, insect repellents, and sunscreen residue. Proper backcountry behavior is critical; visitor education is necessary.

The historic fire regime in the parks has been disrupted by overzealous prevention and suppression of fire over the past several decades. The results include heavy fuel loadings, unnatural species composition, and more frequent diseases in the giant sequoias. A program of prescribed burning has been established to try to recreate the historic role of fire in park ecosystems. Research is being conducted to identify just what the role of fire should be.

SHENANDOAH
Rolling Mountains and Heaven's Hills

Luray, Virginia
ACRES: 199,038

The creation of national parks was an American idea beginning in Yellowstone in 1872. Many parks were added subsequently, mostly in the west. In 1934, Great Smoky Mountains National Park was established and, with Acadia National Park in Maine, was one of only two east of the Great Plains. Concern for lack of protection for park resources in the east led Congress to establish Shenandoah National Park in Virginia. The park was first authorized in 1926, but land acquisition proved costly and time consuming. The park was finally fully established in 1935.

Shenandoah protects 199,038 acres within the Blue Ridge Mountains; 79,579 acres of it are designated wilderness. Skyline Drive, completed in 1939 by the Civilian Conservation Corps, follows the length of the park for 105 miles, providing access to motorists and splendid overlooks to the east and the west. A 101-mile segment of the Appalachian Trail also runs the length of the park; more than five hundred miles of trail exist in Shenandoah. Elevations range from about six hundred feet at the lowest point to the summit of Hawksbill (4,049 feet).

Shenandoah contains several habitats. Forests

Backpacking opportunities abound in Shenandoah National Park, including a 101-mile segment of the Appalachian Trail. The rocky outcroppings provide vistas that include the nearby croplands.

include oak-hickory, hemlock-white pine, oak-pine, and birch-maple. In moist hollows are found cove hardwoods, which include tuliptree, oak, hemlock, basswood, and sometimes maple. Riparian zones are common as well. When the park was established, more than a third of it was grassland and abandoned pasture. Much has returned to forest, but some significant open areas remain, such as Big Meadows along Skyline Drive. This variety in habitat and elevation creates substantial biological diversity; more than fifty species of mammals now inhabit the park.

When Shenandoah was established in 1935, it is believed that the black bear had been extirpated by hunting and by farmers responding to livestock predation. Bears soon returned on their own through migration from adjacent forests. Their population continued to grow; today it is estimated that three to five hundred black bears inhabit Shenandoah.

The bobcat and gray fox are common throughout the park. Coyotes are migrating eastward and their presence in the park has been documented; their populations are likely to increase. The smaller carnivores include the striped skunk and eastern spotted skunk.

Among the rodents, residents include the eastern gray squirrel, southern flying squirrel, and red squirrel. The eastern chipmunk is found throughout the park. Beaver are uncommon, found only in Thornton River. Thornton River is also the park's only home to the muskrat.

Eastern cottontails are found everywhere at all elevations. Other common small mammals include the Virginia opossum, raccoon, and woodchuck.

The most common mammal seen in Shenandoah is the white-tailed deer. All of the park's ungulates—elk, bison, and deer—were extirpated by the time the park was created. Deer were reintroduced several times, the first time in 1934. Their population has grown to several thousand today.

CONSERVATION CONCERNS

Shenandoah's location has provided a multitude of opportunities for the millions of people who live nearby. Unfortunately, that location has created numerous environmental problems for the park. These problems have become so acute that Shenandoah has been on the National Parks Conservation Association's list of America's "Ten Most Endangered Parks" for both 2003 and 2004. The reasons for the listing are described below.

Air quality is one of the critical issues today. Northern Virginia has exhibited substantial economic growth in the recent past. To fuel that growth, new power plants have been built, many of them fueled by coal. Shenandoah is downwind from many of these new plants. Pollutants from more distant sources are also becoming common. The result has been an increase in air pollution in the park and around its borders.

Reduction in air quality includes several major concerns. The average normal distance visible in the park is estimated to be one hundred miles. On the worst days, haze reduces visibility to one mile. Acidification can cause an increase of acids in streamwater, causing adverse effects on aquatic ecosystems. The blacknose dace and the Appalachian brook trout are both acid-sensitive fish; they have already been adversely affected. Deposition also occurs in terrestrial ecosystems. Some of the acid ends up in park waters, while the rest remains in the soil.

Ozone results from the reaction between nitrogen oxides and volatile organic compounds in the presence of heat. In summer, ozone becomes a considerable problem. Some of the park's vegetation is ozone sensitive and damage is already apparent. Ozone also is harmful to human health; there are several days per summer when ambient ozone exceeds human health standards.

In the early part of the nineteenth century, a fungus called the chestnut blight was accidentally introduced into the United States. It was an ecological disaster. The American chestnut once made up 50 percent of eastern mountain forests including Shenandoah; almost all of these trees were destroyed. It has been described as one of the worst natural calamities ever to strike the country.

Unfortunately, history is repeating itself. Another fungus, dogwood anthracnose, is attacking the flowering dogwoods in the park. It kills twigs and branches and creates opportunities for infection to enter the tree. Dogwood is one of the most attractive understory trees in the park, with striking white flowers in the spring. Its loss would certainly reduce the scenic beauty of the park. A second anthracnose on sycamore may also enter the park.

Another tree is being attacked by an alien insect, the hemlock woolly adelgid. It now has destroyed many of the park's hemlocks. Where hemlocks are common, shade can be severely reduced, radically changing an ecosystem. The balsam woolly adelgid has already played havoc with firs in the Appalachians. Are the hemlocks next? Unfortunately, they probably are.

Another foreign insect pest is the gypsy moth. The moth was introduced into the United States in 1869 to improve silk production. Its more lasting legacy has been harm to the east's hardwood forests. The larvae of the gypsy moth feed on several hardwood species including oaks, reducing the trees' vigor and slowing their growth; many ultimately die.

It is estimated that 23 percent of the park's plant species are exotic. Some are well established such as the tree-of-heaven, which covers hundreds of acres in Shenandoah. Kudzu is present in the park but so far it has been kept under control. Totally removing all exotics is probably not possible. Still, management must continue monitoring, preventing new infestations, and seeking ways to control what is already present.

Shenandoah is facing a problem that may become all too common. Most parks are located in the midst of other public lands or are far from urbanized development. Not so at Shenandoah. Development of commercial and residential properties is expanding around the park. Forests and farms get developed; wildlife habitats are destroyed. More highways are constructed, causing further fragmentation of habitat. Management has little or no control over land uses on private lands adjacent to national parks. Some communities have enacted zoning ordinances to control growth; many have not. Ultimately, park managers have only the power of persuasion.

THEODORE ROOSEVELT

A Monument to the Conservation President

Medora, North Dakota

ACRES: 70,447

O f the fifty-six national parks in the United States, most are named after a geographical feature (Grand Canyon, Yosemite) or unique vegetation (Redwood, Joshua Tree); only one is named for a person. Theodore Roosevelt probably had more to do with natural resource conservation than any other president. Roosevelt was president when the first national wildlife refuge was created, the national forest system was expanded, five national parks were established, and the Antiquities Act was passed, which gave presidents the authority to declare sites as national monuments. Roosevelt declared the first eighteen monuments including the very first, Devils Tower National Monument in Wyoming.

Theodore Roosevelt National Park is one of several National Park Service sites to commemorate his life and presidency. Roosevelt first visited the area in 1883 on a hunting expedition where he saw the effects of hunting on big game populations and the results of grazing on rangelands. In 1884 Roosevelt returned to North Dakota and established Elkhorn Ranch. He was already part owner of the nearby Maltese Cross Ranch. His time in the North Dakota badlands established in Roosevelt's mind the need for conservation. He later wrote, "I never would have been President if it had not been for my experiences in North Dakota."

Theodore Roosevelt National Park consists of a North Unit (46,000 acres) and South Unit (24,000 acres) with the 218-acre Elkhorn Ranch Site in between. Total park acreage is 70,447 acres, 29,920 of which are designated wilderness. The primary landform is the badlands, the result of water and wind erosion. Most of the park is grassland with some sagebrush. In riparian areas are scattered stands of willow and cottonwood; some juniper grows in uplands. Both units have scenic drives that allow visitors to enjoy the park's ecosystems and wildlife.

Probably the most common carnivore is the coyote. Their songs usually begin at nightfall, lending to the aura of the wide open prairie. The other two park canids are the red fox and the swift fox.

The felines are rarely seen. The one most often seen is the bobcat, but only a few sightings are reported each year. The lynx and mountain lion are both very rare at Theodore Roosevelt.

The smaller carnivores are mostly members of the weasel family. Badgers are most often encountered near prairie dog colonies. The striped skunk is quite common; the long-tailed weasel is rare.

In the riparian areas, beaver are probably the most common mammal. They may be seen along the Little Missouri River and Squaw Creek in the North Unit. Mink and muskrats are sometimes seen. Raccoons are sometimes seen in the evenings but are uncommon.

Among the lagomorphs, desert and eastern cottontails may be observed, but Nuttall's cottontail is very rare. In open country, the white-tailed jackrabbit (prairie hare) is sometimes seen.

The most frequently observed rodent is the black-tailed prairie dog. It is actually related to chipmunks and ground squirrels. The scenic drive in the

Many of Theodore Roosevelt's views about natural resource conservation were formed during his stay at the Maltese Cross Ranch in North Dakota. Today his namesake park is one of the sites where bison have rebounded, with virtually every visitor seeing a herd during his or her stay.

(Pages 84–85)
Star tracks appear
behind giant
sequoias in Sequoia–
Kings Canyon
National Park.

South Unit passes through three large prairie dog towns. The least chipmunk is seen everywhere; the thirteen-lined ground squirrel is abundant and frequently encountered by visitors.

Ungulates reside in the park; some of them have been reintroduced. Mule deer prefer the park's uplands; during the heat of the day they bed down in juniper groves. White-tailed deer prefer the woodlands and riparian areas but many times are found in campgrounds and picnic areas. Pronghorns live in the open country in both Units. Elk were extirpated in the park but were reintroduced in 1985. They now are sometimes seen in the South Unit grazing in open grasslands at dusk and dawn.

In 1883 Theodore Roosevelt hoped to successfully hunt a bison, but by then most were gone. It is thought they were extirpated in the region in 1884 but some may have survived. A federal law was passed in 1894 protecting bison but only about three hundred remained in the country. Bison raised on game preserves allowed the population to increase.

In 1956, twenty-nine bison were imported from Fort Niobrara National Wildlife Refuge in Nebraska and placed in the South Unit. Three years later, some of the bison were transferred to the North Unit. To maintain a healthy range, herds in the North Unit are limited to one to three hundred bison; in the South Unit the limit is two to four hundred. Care must be taken not to get too close to them, as some have referred to the bison as "the most ferocious animal in North America."

The historic population of bighorn sheep, the Audubon bighorn, was extirpated early in the twentieth century. In 1956 bighorns were reintroduced in the South Unit and seem to be doing well. Look for them near Peaceful Valley and the Beef Corral along the scenic drive in the South Unit.

Theodore Roosevelt National Park is somewhat unusual in that herds of two feral animals are maintained as a demonstration for park visitors. Prior to 1970 the park's policy was to remove all wild horses in the park. In 1971, Congress enacted the Wild and Free-Roaming Horses and Burros Act giving protection to these animals as "national heritage species." Theodore Roosevelt National Park maintained management authority of wild horses in the park, and in 1970 changed their policy to one of retention, demonstrating the role of the horse in the park's historical setting. Today, a herd of 70–110 animals is retained in the South Unit. Excess horses are rounded up and sold at auction. Look for them along the park road in the Peaceful Valley area.

Likewise, a small herd of longhorn cattle is maintained in the North Unit. They may be seen at the Longhorn Lookout. During the heat of the day, they may be found in the cottonwoods along streams.

CONSERVATION CONCERNS

Theodore Roosevelt National Park is located far from large metropolitan areas. Urbanization has not been a problem to date. Air and water quality can be harmed by deposition of pollutants carried aloft for long distances. To date, no serious harm has been seen at Theodore Roosevelt.

One serious problem, however, has been the introduction of exotic plants. More than sixty exotic species have been identified to date. Control efforts have been aimed at leafy spurge, spotted knapweed, and Canada thistle. Chemical, mechanical, and biological control methods are being used. This is likely to be a long-term effort.

VIRGIN ISLANDS
Bays of Crystal Blue-Green Waters

St. John, Virgin Islands
ACRES: 14,689

The only national park outside the fifty states, Virgin Islands National Park offers a much different kind of ecosystem than those found in other parks. The park, on the island of St. John, has seen huge changes over time. Most park lands are former sugar plantations, cotton fields, and pastures. Many of the park's plants were introduced for food and medicine; most of the original forest is gone.

Virgin Islands National Park was donated to the nation by Laurance Rockefeller in 1956. Total park acreage is 14,689. The park occupies a little more than half of St. John; the rest of the island is privately owned homes, businesses, and undeveloped lands. Much of the offshore waters around St. John are within the park boundary.

Even though it is a small park, Virgin Islands attracts significant visitation; 769,962 visits were recorded in 2003. Attractions include extraordinary beaches and coral reefs. Trunk Bay even has an underwater interpretive trail illustrating several kinds of marine creatures. Park waters offer sailing and fishing as well as premier diving and snorkeling opportunities. The land portion is characterized by deep valleys; 86 percent of the island's surface is on slopes greater than 20 percent. Hiking ranges from easy to difficult. Cultural resources abound.

Wildlife in the park consists mostly of birds, reptiles, and fish. Two endangered sea turtles inhabit the waters of St. John, the hawksbill and green turtles. The only native mammals in the park are five species of bats: Antillean fruit bat, greater bulldog bat, Jamaican fruit bat, Palla's mastiff bat, and red fruit bat. Some are quite common; during the day the number of bats active may exceed the number of birds.

Numerous species of whales may be seen in nearby waters. The West Indian manatee was recorded at West End, Tortola, in the British Virgin Islands. The site is only two to three miles from Virgin Islands National Park. A recent archeological dig found the remains of two extirpated mammals, the Caribbean monk seal and the Puerto Rican shrew. The other mammals in the park are exotics and are described below.

CONSERVATION CONCERNS

In 2003 Virgin Islands National Park was listed as one of America's "Ten Most Endangered Parks" by the National Parks Conservation Association (NPCA). The reasons for the listing dealt with overfishing and destruction of corals. Since no-take zones were being enforced more vigorously, NPCA removed Virgin Islands from the list for 2004.

Though there are few native mammals in Virgin Islands, there are several feral species and exotics. Feral species include hogs, goats, sheep, cows, dogs, house mice, and donkeys. The exotics include the small Asian mongoose, wild cats, and the white-tailed deer. Norway and black, or tree, rats also made their way to St. John.

Hogs are not native to the Americas. They were introduced to the West Indies by Christopher Columbus in 1493; the Danes also brought them to St. John during colonization in the 1700s. They have become feral and have spread throughout St. John and the park. The rooting of hogs disrupts the park's native communities; they are especially affecting reptiles and invertebrates. Their rooting on trails causes increased erosion. The plan now is for a sustained reduction in hog numbers. Traps and fencing will be used initially; shooting and dogs will be used for elusive hogs.

Feral goats and sheep arrived in the Virgin Islands by the same means as the hogs. Their trampling and grazing behavior has seriously harmed the park's natural ecosystems. They especially harm fauna that are supported by fruits and berries, in-

cluding reptiles, amphibians, and insects. An estimate of the number of feral goats on the island is between six hundred and a thousand; sheep number about fifty. These numbers include ferals that live in the park or graze in it periodically. A draft plan for removal of feral goats and sheep was released in December 2003. Livestock owners will have to remove their animals from the park. They then must be fenced, registered, and visibly tagged. Those in violation will be removed.

A similar program will be put in place for rats, cats, and mongooses. The cats on the island are descendants of European and African wild cats. The mongoose was introduced to prey on tree rats in the sugarcane fields. Unfortunately, the tree rats are nocturnal and spend the day in trees, while the mongoose is out during the day and cannot climb trees. Their prey became ground-nesting birds and reptiles. A plan to reduce the population of these exotics in a sustainable manner is being put into place. The goal is to reduce populations to a point where they don't harm native communities.

Donkeys have a more direct effect on visitation. Their travels around the island cause severe erosion on trails. They frequently enter campgrounds, destroy tents and equipment, and have been known to bite visitors.

Several species of threatened and endangered plants are being harmed by the feral and exotic fauna. The controls being used are certainly necessary. Hopefully, they will be successful.

Visitation has grown ten-fold in the past decade. Much of the growth relates to the cruise ship industry. With the cruise industry building much larger ships, pressure in Virgin Islands is expected to grow. It will mean many more visitors in short intervals of time. Crowding is a likely outcome.

America's national parks extended deep into the Caribbean with the addition of Virgin Islands National Park in 1956. The only national park outside the fifty states offers blue-green waters for incredible snorkeling opportunities.

VOYAGEURS

A Canoeist's Paradise

International Falls, Minnesota
ACRES: 218,200

The life of the voyageur was certainly exciting. For hundreds of years these French-Canadian boatmen would depart Montreal in birchbark canoes laden with trade goods that would eventually be exchanged for furs. Their route took them through lakes and rivers of Quebec and Ontario, then across the length of Lake Superior. Most crossed the Grand Portage in Minnesota, finally passing through the Rainy Lake country to their destination in the foothills of the Rocky Mountains. The route became so well known that in 1783 when the boundary between the United States and Canada was established, the "voyageurs highway" was used between Minnesota and Ontario.

Each canoe was manned by twelve to fifteen boatmen. Cargo was packed in packages of up to ninety pounds each. On portages, an individual voyageur would carry as much as two hundred pounds of freight, frequently making several trips. They carried all their own supplies as well. Meals consisted of peas, salt pork, and biscuits. After paddling for sixteen hours per day, they slept under their overturned canoes. Each trip lasted several months and was fraught with danger, yet the lifestyle attracted many.

A fifty-six-mile segment of the "voyageurs highway" was preserved in Voyageurs National Park in 1975. Rainy Lake is the largest lake in the park and forms its northern boundary as well as the international boundary. In total, the park contains 218,200 acres of lakes and forests; about one-third of the park is water.

Voyageurs is primarily a wilderness park. There is vehicle access to Kabetogama Lake and Rainy Lake as well as to Ash River and Crane Lake. Visits to the interior of the park involve travel by water, including motorboats, houseboats, sailboats, kayaks, and canoes. Water-accessible backcountry campsites are provided on lakes large and small. A concessionaire-operated tour boat departs from Rainy Lake Visitor Center for cruises on Rainy Lake. In winter, the seven-mile long Rainy Lake Ice Road allows cars to cross the ice to the Kabetogama Peninsula and out along Rainy Lake. Snowmobiles also provide access to the park's interior; several snowmobile trails are maintained. Visitation in 2003 was 237,447, primarily boaters and fishermen.

Voyageurs is located in the southern part of the Canadian Shield, some of the oldest exposed rocks in the world. Glaciers removed later deposits, leaving evidence of their passing as scrape marks on the 2.7 billion-year-old rock beneath. Depressions left by the glaciers became ponds, lakes, and streams. The northern boreal forest soon covered the landscape, containing pine, spruce, aspen, birch, and maple. The mammals in the park are those typical of this forest type. Due to the park's water resources, aquatic mammals are very common.

Gray wolves travel and hunt in packs throughout the park. In winter, they use frozen lakes and ponds as easier travel routes; travelers present then may see wolves since they are out during the day in winter. Wherever wolves are common, coyotes are scarce. The red fox is common in Voyageurs but seldom seen.

Black bears are active during most of the day. Bear-proof storage boxes are provided at backcountry campsites. Among the felids, lynx and bobcat are probably present, but their status is unknown.

The smaller carnivores are common as well.

Almost as remote as an Alaskan park, Voyageurs offers little access to those without a canoe. Those with a canoe are rewarded with peace and quiet as they follow the route of the French-Canadian voyageurs in Minnesota's North Country.

A fast-moving storm brings moisture to the prairie above South Dakota's Wind Cave. This national park was established by Theodore Roosevelt and today protects both the extensive cave system and one of America's most beautiful prairies.

These include the striped skunk, river otter, mink, and short-tailed weasel (ermine). Pine martens and fishers are uncommon. The long-tailed weasel is probably present, but its status is unknown.

The red squirrel is the park's most common tree squirrel. Another is the northern flying squirrel. They are uncommon and are extremely secretive. There are two chipmunk species in the park, the eastern chipmunk and least chipmunk. Franklin's ground squirrel is also present. The thirteen-lined ground squirrel is an unconfirmed resident of the park.

The park's extensive water bodies attract a large population of beaver. Beaver at Voyageurs may be active during daylight. The same habitats that support beavers also support muskrats, though they are considered uncommon in the park.

Both white-tailed deer and moose are common. Moose are active at night but are sometimes seen during daylight.

CONSERVATION CONCERNS

To date, the air and water of Voyageurs National Park remains nearly pristine. No urban areas exist nearby, and visitation remains low compared to other parks. The large paper mill in nearby International Falls, Minnesota, controls pollutants. Acid deposition may become a problem from long-distance sources; monitoring of terrestrial and aquatic ecosystems must continue. Long term, global warming could have a serious effect on the park, but the ramifications are now unclear.

WIND CAVE
Honeycomb Catacombs beneath Bison Prairie

Hot Springs, South Dakota

ACRES: 28,295

In the late nineteenth century, stories existed about holes blowing wind in the Black Hills of South Dakota. In 1881 Jesse and Tom Bingham were attracted to a hole because of a whistling noise. When Tom looked into the hole, his hat blew off, hence the cave's name, or so the story goes. Mining claims were filed that included the cave. Wind Cave quickly became a tourist attraction managed by private interests. Competitors soon arrived on the scene and conflicts abounded. The Department of the Interior later declared that since no mining had taken place, there would be no valid claims and any land claims on the cave would be returned to federal ownership. President Theodore Roosevelt signed a bill two years later in 1903 establishing Wind Cave National Park.

Initially there was little wildlife in the park; the cave was its only focus. In 1912, a national game preserve was established bordering Wind Cave managed by the U.S. Biological Survey, forerunner of the Fish and Wildlife Service. In 1935, the park and game reserve were merged under National Park Service management. Total acreage of the new park was 28,295 acres.

Cave tours remain popular in the park; 842,801 visitors entered the park in 2003. But the surface provides ample opportunity to enjoy Black Hills wildlife. The park has multiple habitats including mixed-grass prairie, ponderosa pine forests, and riparian woodlands containing ash, elm, bur oak, and box elder. Grassland is the most common habitat, covering 75 percent of the park. Elevations in the park range from 3,650 feet to the Rankin Ridge Lookout at 5,013 feet. The variety of habitats makes Wind Cave an excellent place to see mammals of the prairie and ponderosa pine forest.

Mountain lions exist in the park; bobcats are rarely seen while the coyote is seen everywhere. Badgers are occasionally seen in prairie dog towns. Other resident small carnivores include the raccoon, short-tailed weasel (ermine), striped skunk, and long-tailed weasel.

Two species of cottontails inhabit the park including the eastern cottontail and the desert cottontail. Both are frequently seen grazing on the lawn of the visitor center. The white-tailed jackrabbit is also present.

Rodents are extremely common at Wind Cave. The most sought-after rodent is the black-tailed prairie dog. There are several prairie dog towns in the park. The black-footed ferret historically has been a primary predator of prairie dogs and was frequently found around prairie dog towns. Unfortunately, they have been extirpated from the park, last seen here in 1971. It is believed that there are too few prairie dogs to support a ferret population and that there are too many roads within prairie dog habitat.

Other common rodents include the least chipmunk, thirteen-lined ground squirrel, red squirrel, and porcupine. The bushy-tailed woodrat, or pack rat, lives in rocky areas and around the cave entrance. Sometimes it enters the cave.

In the late nineteenth century, bison, elk, and pronghorn were extirpated from the Black Hills due to overzealous hunting. When Wind Cave was established, none of the three existed in the park. In 1912 the American Bison Society sought a site to re-establish a bison herd; the following year they chose the game preserve surrounding Wind Cave due to its excellent habitat. Elk and pronghorn were reintroduced about the same time. Today bison and elk are abundant, so much so that their herds must be culled to maintain range quality.

There are about 350 bison in the park. They are commonly seen along park roads and around the campground. Elk are abundant in the park. Pronghorns are resident in grasslands but are uncommon. Other ungulates in the park are the mule deer and white-tailed deer. Both are frequently observed.

CONSERVATION CONCERNS

As noted above, the park lacks some important parts of the historical fauna. The black-footed ferret is gone and reintroduction is unlikely. The park contains populations of large ungulates, but except for a few mountain lions, the large predators have been extirpated. As bison and elk populations grow without predation, there is concern for the quality of habitat. Bison and elk have had to be reduced, with most of the excess animals being moved to other parks and reservation lands.

Another biological concern in the park is exotic flora. More than 20 percent of the plants in Wind Cave are exotics. Many create monocultures that destroy diversity. Solutions include biological, mechanical, and chemical control. Finding the correct combination is a long-term effort.

In 1997 a mule deer suffering from chronic wasting disease (CWD) was found on land adjacent to the park. CWD is a fatal disease affecting the brain and central nervous system of deer and elk. A study of CWD is now underway to assess possible future impact on herds. To date, the disease has not been found inside Wind Cave, but the potential harm to two important park mammalian species is extreme.

Concerns about air and water pollution have resulted in monitoring programs being established in the park. To date, no significant harm to air quality has been found, but diligence is required to identify future impacts, especially from long-distance deposition.

In 1996 a dye-trace analysis of cave hydrology was undertaken. Dyes were inserted in two places, but ended up in the cave at the same place. Results show that seepage is not necessarily vertical and that horizontal movement is quite likely. The visitor center with its sewage system was built directly above the cave. Seepage of sewage into the cave has not been shown to date, but maintenance of the sewer system is critical.

Surface land uses may also have negative impacts on cave ecosystems. Parking lots and roads collect small amounts of gasoline, antifreeze, motor oil, and lubricants that can move through the soil into the cave. Asphalt particles from the roadways can also enter the cave; residues from other human activities may also have an impact. Efforts are now under way to study how such seepage can be prevented. Cave ecosystems are very fragile; even minute amounts of pollutants can cause harm.

WRANGELL–ST. ELIAS

The Mountain Kingdom of North America

Cooper Center, Alaska

ACRES: 13,175,901

Alaska is a land of superlatives. Not only is it America's largest state, it also contains the most national park acreage of any of the fifty states. Among those national parks is America's largest, Wrangell–St. Elias National Park, containing 13,175,901 acres.

Wrangell–St. Elias is the site where three mountain ranges come together, the Chugach, Wrangell, and St. Elias Mountains. The three contain the continent's largest collection of glaciers and peaks higher than sixteen thousand feet, topped by Mount St. Elias, the second tallest peak in the United States at 18,008 feet. Some of the glaciers are extraordinary; Malaspina Glacier is larger than the state of Rhode Island.

Wrangell–St. Elias was first developed by mining interests. Production of copper at the Kennecott Mine began in 1911 and continued until the rich ore was exhausted in 1938. Alaska's last gold rush occurred in the park's Chisana area in 1913. In succeeding years, attempts to develop tourism met with little success.

In 1971 the Alaska Native Claims Settlement Act authorized the withdrawal and study of federal lands in Alaska for future uses as national parks, forests, and refuges. President Jimmy Carter declared the area a national monument in 1978 to protect its scientific and cultural significance. Finally, in 1980 Congress passed the Alaska National Interest Lands Conservation Act (ANILCA), which included the designation of Wrangell–St. Elias as a national park. National park status was given to 8.3 million acres; the remainder, 4.8 million acres, was designated as Wrangell–St. Elias National Preserve. In the preserve, sport hunting and trapping are allowed.

In addition, Wrangell–St. Elias has received designation by the United Nations as an international World Heritage Site. The site originally consisted of Wrangell–St. Elias and Canada's Kluane National Park in the Yukon Territory. In 1993, Glacier Bay National Park and Alsek-Tatshenshini Provincial Park in British Columbia were added. These four areas contain twenty-four million acres of contiguous protected lands, among the largest in the world.

Wrangell–St. Elias is affected by two main climatic zones—coastal maritime and subarctic continental. The coastal maritime zone is influenced by the sea and regular precipitation and is very moist. The subarctic continental zone is characterized by low moisture, moderate precipitation, and widely fluctuating temperatures.

Vegetation is a function of the climate and elevation. The interior portions are mostly above timberline—39 percent of the park is alpine and sub-alpine tundra. Along river corridors are forests of aspen, birch, and white spruce. Coastal forests of Sitka spruce and interior taiga (Russian for "land of little sticks") forests of white and black spruce provide a variety of habitats. Forests cover 8 percent of the park, while wetlands exist in 1 percent. More than half the park is covered by permanent ice and snow.

Wrangell–St. Elias is prime habitat for the grizzly bear. It is found throughout the park. Black bears are widespread. Wolves are present in the park but are seen only infrequently.

The ungulates are very common in the park. Caribou are found throughout the park in migrating herds. The population of the Mentasta herd has declined to a point to cause concern among park management staff. Moose are found throughout the park. Dall's sheep are found in high elevations where they can scan the landscape for predators. It is estimated that thirteen thousand of them inhabit the park, the continent's largest concentration.

The mountain goat can be seen at high elevations. The Sitka black-tailed deer (a subspecies of mule deer) spends the summer at higher elevations, moving lower in autumn and remaining until spring. There are two herds of introduced bison in the park.

Little is known of the park's furbearer population. These include members of the weasel family: wolverine, short-tailed weasel (ermine), river otter, mink, and marten. Of the canids, the red fox and the coyote are present. The only felid in the park is the lynx.

Numerous smaller herbivorous mammals support the predators. The snowshoe hare inhabits the park, as does the collared pika. Porcupines are sometimes seen in forested areas.

Wrangell–St. Elias is America's largest national park, has the nation's largest glacier, and contains North America's second largest peak, Mount St. Elias, which rises to 18,008 feet.

Several rodents can be found in the park. The Arctic ground squirrel can be found almost anywhere in the park. In forested areas, the red squirrel is resident as is the northern flying squirrel. In higher elevations, you may find a hoary marmot. In riparian areas, beaver and muskrat are sometimes seen.

Wrangell–St. Elias National Park includes 125 miles of coastline on the Gulf of Alaska. There are two major inlets, Icy Bay and Disenchantment Bay. These waters contain several species of marine mammals including sea otters, harbor seals, and the federally threatened northern (Steller) sea lion. Offshore are Dall's porpoise and killer whale, or orca.

CONSERVATION CONCERNS

In most cases, remoteness helps protect resources. Such is not the case at Wrangell–St. Elias. In fact, the park is included on the "Ten Most Endangered Parks" list, compiled by the National Parks Conservation Association (NPCA). There are two reasons for the designation. First, the park authorizes the use of all-terrain vehicles (ATVs) on thirteen trails within the park. Irresponsible ATV use has resulted in damage and loss of vegetation, exposure of permafrost, and creation of extremely muddy areas. These trails are also used by hikers. The result is a growing negative impact along these trail routes.

The second concern results from a Bush administration decision in 2002 to revisit a Civil War–era mining law, RS 2477. This law would allow states to identify potential road segments to be developed on federal land. The state of Alaska identified 1,702 miles on ninety-six routes in Wrangell–St. Elias, half of all identified in Alaska's national parks. The impacts of road building in this extraordinary wilderness could be catastrophic, including noise, habitat fragmentation, and possibly poaching.

The passage of ANILCA authorized subsistence uses in several of the new national parks, including Wrangell–St. Elias. Subsistence refers to traditional uses of resources, such as fishing, hunting, and trapping within the park by local rural residents. Monitoring the impact of subsistence use on mammal populations is critical.

Cruise ship activity has been increasing in Icy Bay. Icy Bay is habitat for the northern (Steller) sea lion, a threatened species. Its impact on them is unknown but will have to be examined in the near future.

Visitation at Wrangell–St. Elias is low; 2003 visitation was 43,311. However, potential for growth is high, especially when cruise ship numbers are included. Additional road construction would likely cause visitor numbers to soar. Protecting the wilderness character of the park may become more and more difficult.

YELLOWSTONE
America's First National Park

Yellowstone National Park, Wyoming
ACRES: 2,219,791

The story of Yellowstone is an interesting tale about a place where just enough was known to pique the interest of those who would follow. The result was the creation of the world's first national park. The first American of European descent to visit Yellowstone was thought to be John Colter, a member of the Lewis and Clark expedition. Colter had returned east with Corps of Discovery but wanderlust drew him back west. For three years beginning in 1806, Colter roamed the area that would ultimately become Yellowstone. His specific route, however, is unknown.

The stories of the incredible scenery spread; the mountain men probably were not interested in thermal features, but they were certainly interested in furbearing mammals. Many of them came to Yellowstone until the fur trade died out in the 1830s. Prospectors soon followed, looking unsuccessfully for gold. By the 1860s, many people had heard of Yellowstone, but few really knew what it contained.

Beginning in 1869, expeditions entered Yellowstone to describe its attributes, culminating in the Hayden Expedition of 1871, led by Dr. Ferdinand Hayden of the U.S. Army Topographical Engineers. The descriptions of Yellowstone's geysers, mud pots, fumaroles, and other features created public and political pressure to protect these resources. In 1872, Congress created Yellowstone National Park, the world's first national park.

In the early years, Yellowstone was a wilderness with few visitors. There was little money or staff to protect the park, especially from poachers. In 1886, to protect the park more effectively, responsibility for management was transferred to the U.S. Army, which managed Yellowstone until the National Park Service was established in 1916. Development quickly followed after 1916 as the first director of the Park Service, Stephen T. Mather, supported the building of roads, hotels, and campgrounds. Use continued to grow and by 2003, reached 2,995,640 visitors.

Yellowstone is a huge park, bigger than Rhode Island and Delaware combined. Current acreage is 2,219,791. The park is primarily forested (80 percent) with grassland covering 15 percent and water 5 percent. Lodgepole pine is the most common tree in the park. It is the pioneer species after a fire, covering huge acreages burned in the 1988 wildfire.

Elevations range from 5,282 feet along Reese Creek to the 11,358-foot summit of Eagle Peak. As a result, the park contains a wide variety of plants and animals including about sixty species of mammals.

One of the reasons people come to Yellowstone is the chance to see a bear. In the past, bears were seen frequently along highways, causing the infamous "bearjams." Garbage dumps attracted grizzly bears; campground bears sought food from people nightly. Even with visitation in the millions, however, only five fatalities related to bears have ever occurred in the park. Most injuries were caused by black bears, most often while people were trying to feed them or get a picture.

In the 1980s an attempt was made to make bear behavior more natural by restricting human behavior. A ban on feeding wildlife, including bears, was strictly enforced. Food storage requirements were also enforced; no coolers or food sources can be left outside a tent or vehicle at night. As a result, few bears are seen in campgrounds or along roads.

There are between 350 and 400 grizzly bears in the park. Black bears are more common in the park than grizzlies. The estimated population today is between five and six hundred. Black bears are resident in forests and meadows.

Wolves had been resident in the park but were caught up in the philosophy of removing predators in the 1920s. The last reported wolf killed in the park was in 1926. In March of 1995, fourteen wolves were released into the wilds of Yellowstone. These wolves had been brought to the park from Canada; more were released in 1996. They multiplied and spread throughout the Yellowstone ecosystem. A recent estimate on the number of wolves in Yellowstone is 162. Consult park staff to see where current wolf activity may be.

Coyotes are still common in the park in forests, meadows, and grasslands. Scientists now believe that coyotes and wolves will learn to coexist just as they did before humans removed the wolves. The mountain lion was severely hunted because of its role as a predator, but they were probably not all removed. Since the end of predator control, the park's lion population has slowly grown. Today, it is estimated that eighteen to twenty-four lions reside in the park.

The weasel family is very common at Yellowstone, including the pine marten, long-tailed weasel, and short-tailed weasel, or ermine. River otters are also common in the park's water bodies including Yellowstone Lake.

Among the park's lagomorphs, the most common is the snowshoe hare. The desert cottontail and the mountain (or Nuttall's) cottontail are both present.

Many rodents are seen by visitors. There are three species of chipmunk: the least, Uinta, and yellow pine chipmunk. Other rodents include the yellow-bellied marmot, golden-mantled ground squirrel, Uinta ground squirrel, red squirrel, northern flying squirrel, porcupine, and bushy-tailed woodrat, or pack rat.

The population of Yellowstone's beavers is between 300 and 350. Muskrats usually are commonly found in the same locales as beaver.

Elk are the most abundant ungulate in the park. When Yellowstone was established, market hunting of the park's ungulates was widespread. The presence of the army after 1886 stopped the slaughter and allowed elk populations to rise. Today, elk numbers in Yellowstone are estimated

The long rays of the sun are reflected in a backcountry lake against Bunsen Peak in Yellowstone National Park.

at thirty-five thousand in summer.

The largest mammal in the park is the bison. The current population is between twenty-two and twenty-five hundred. Bison in Yellowstone are unique in that they are the only continuously wild population since prehistoric times in the United States.

The moose is the largest member of the deer family and is commonly seen in Yellowstone. Moose are found in riparian areas during summer typically feeding on aquatic vegetation.

Yellowstone is home to two deer species. Mule deer are found in most habitats almost anywhere in the park. Their estimated population in Yellowstone is twenty-five hundred. The other is the white-tailed deer. They are found in all park habitats but are rare.

Bighorn sheep used to be very common in the west, but their numbers were drastically reduced by hunting. Bighorn populations have not reached pre-historic levels, numbering 150 and 225 today. They spend summer in alpine meadows or along cliff faces. Pronghorns are mammals of grassland and sagebrush. There are between 200 and 250 of them in Yellowstone.

Finally, a few mountain goats have found their way into the park. The mountain goats are not native but were introduced. Their preferred habitat is rocky slopes.

CONSERVATION CONCERNS

Yellowstone National Park receives nearly three million visits annually, but surprisingly the most contentious issues in the park involve bison and snowmobiles. These issues have been so serious that Yellowstone remains on the National Parks Conservation Association's (NPCA) list of America's "Ten Most Endangered Parks" for 2003 and 2004.

The issue regarding bison concerns a bacterial disease called brucellosis. Brucellosis was introduced into the United States by European cattle; the main concern is that it can cause cattle to miscarry. About half of the park's bison are carriers of brucellosis. However, there is no record of wild bison ever transmitting the disease to cattle or humans. The state of Montana has a "brucellosis-free status" that allows Montana ranchers to ship their cattle out of state without testing. The Montana Department of Livestock and the livestock industry are concerned about losing that status if cattle come into proximity with bison. Therefore, when bison leave the park, especially in winter to seek additional food, they may be shot.

The National Park Service banned snowmobiles from the park due to noise and air pollution. The exhaust from snowmobile two-cycle engines greatly reduces air quality. Park employees staffing fee booths must wear masks in order to breathe. Opponents point to the positive economic impact snowmobiles bring to the winter economy. They also note that four-cycle engines are now available. The ultimate outcome remains to be seen.

Air quality also may be harmed in the park if a proposed power plant is built nearby. Yellowstone has been fortunate in that there is no large metropolitan area nearby. However, pollutants are traveling ever farther, and combined with power plant emissions air quality in the park could suffer.

Sometime in the last thirty years, someone introduced lake trout into Yellowstone Lake. One lake trout can consume up to sixty native cutthroat trout per year, a species that supports many wildlife species including the grizzly bear. Lake trout are not a substitute because they spawn in deep water where bears can't get them. Park staff is trying to gill net lake trout, and incentives have been given to anglers to catch them. It is hoped that these measures will be successful in controlling lake trout numbers.

One potential issue concerns thermal features. There have been proposals in the past by landowners outside the park to tap into the geothermal energy in the area. Where this has been done elsewhere in the world, thermal features have been harmed or destroyed. As Americans cast a wider net for energy supplies, we must consider the ramifications of any new energy development.

YOSEMITE
Waterfalls and Mountain Streams

Yosemite National Park, California
ACRES: 761,266

Most people agree that Yellowstone was the world's first national park. There are, however, a few dissenters. Soon after the California Gold Rush, entrepreneurs saw the possibilities of tourism in Yosemite Valley. Hotels, residences, orchards, and livestock grazing were established, and as a result the valley's natural character suffered greatly. California's Senator John Conness appealed to President Abraham Lincoln for help. On June 30, 1864, Lincoln took time away from wartime concerns to sign a bill granting to the state of California, Yosemite Valley and the Mariposa Grove of Giant Sequoias for state management as a park. This was the first time the federal government had

A windless summer day enhances the reflection of this granite cliff in Yosemite National Park.

set aside lands for public enjoyment and preservation. It became the precedent used in 1872 to create Yellowstone National Park.

The foothills surrounding Yosemite Valley continued to be developed. Through the efforts of John Muir and others, Yosemite National Park was established in 1890, protecting the lands around the state park. For a time federal and state park management existed side by side. However, in 1906, California returned the state lands in Yosemite to the federal government, creating today's Yosemite. Current park acreage is 761,266.

John Muir first arrived in Yosemite in 1868. His writings made the park known to the rest of the country and the world. Following is a Muir description:

No temple made with hands can compare with Yosemite. Every rock in its walls seems to glow with life. Some lean back in majestic repose; others, absolutely sheer or nearly so for thousands of feet, advanced beyond their companions in thoughtful attitudes, giving welcome to storms and calms alike, seemingly conscious, yet heedless of everything going on about them. Awful in stern, immovable majesty, how softly these mountain rocks are adorned and how fine and reassuring the company they keep.

The granite Muir describes forms the high country up to the summit of Mount Lyell (13,114 feet). Visitors can get to the high country on the Tioga Pass Road, which reaches 9,945 feet. An extensive trail system, including a segment of the Pacific Crest Trail, provides access to the backcountry. Elevations range down to two thousand feet.

The elevation difference and moisture patterns from the Pacific create a variety of habitats. On the wetter west side of the park, forests exist below eight thousand feet, with oak woodland at the lowest elevations and mixed conifers up to the subalpine area.

On the drier east side, lowest elevations are sagebrush covered interspersed with pinyon pine. An alpine area exists above timberline at ten thousand feet. With all the diversity of habitat, approximately ninety species of mammals live in the park.

Black bears are common in the park; it is estimated that their population is from three to five hundred. The park's other large predator is the mountain lion. Coyotes may be the most commonly seen predator. The smaller predators are mostly members of the weasel family. They include the short-tailed weasel (ermine), long-tailed weasel, striped skunk, and western spotted skunk.

The largest group of mammals in the park is the rodents with thirty-nine species. The most abundant is the California ground squirrel. Three other ground squirrels are common, the golden-mantled ground squirrel, Belding's ground squirrel, and the yellow-bellied marmot.

Yosemite has the most species of chipmunk of any national park—eight. The most common of the chipmunks are Allen's, Merriam's, lodgepole, and yellow-pine chipmunks. There are three forest-dwelling squirrels in Yosemite, the western gray squirrel, Douglas's squirrel, or chickaree, and the northern flying squirrel. There are two species of woodrat in the park, the dusky-footed woodrat and the bushy-tailed woodrat.

Lagomorphs in Yosemite include the brush rabbit, black-tailed jackrabbit, and white-tailed jackrabbit. Ungulates found in Yosemite are mule deer and mountain bighorn sheep. The latter is included on California's threatened species list and is being considered for the federal list.

CONSERVATION CONCERNS

Yosemite is one of America's most popular parks; 2003 visitation was 3,380,038 visitors. Most visitors flock to Yosemite Valley, while places like Hetch Hetchy are little used. To avoid crowding and traffic congestion, Yosemite has instituted a shuttle system to help remove traffic from Yosemite Valley. Shuttles move visitors throughout Yosemite Valley, with cars parked near park entrances. At times in the past, park gates have been closed when Yosemite Valley was considered full. It is hoped that such action will not be necessary in the future.

With such high visitation, water and air quality issues could become apparent. However, water quality throughout the park is considered to be good and generally above state and federal standards. In some parts of the park, water quality is regarded as pristine; in some heavily used areas, water quality has been somewhat degraded.

A much bigger concern is air quality, since sources of pollution can be worldwide. A monitoring program is in place to determine the severity of air quality issues. Regarding acid deposition, chronic problems do not occur at present, but acidification sometimes occurs when loadings are high, due to the fact that the park's natural buffering is somewhat low.

Ozone pollution occurs when nitrous oxides are heated during summer. Excessive ozone can harm vegetation as well as human health, especially among people with respiratory problems. Ozone levels exceed the unhealthy level in Yosemite approximately five times per summer. Because of considerable growth in industry, agriculture, and use of motor vehicles in the nearby San Joaquin Valley, episodes of unhealthy ozone levels are expected to increase.

Yosemite is known for its spectacular scenery, but reduced visibility is causing some of the park's major sights to be clouded, including Half Dome, El Capitan, and Yosemite Falls. Haze is a function of particulates, which may come from dust, fires, and diesel-powered vehicles. Particulates cause more absorption and scattering of light, reducing visibility.

Another potential airborne problem is pesticides, which are widely used in the San Joaquin Valley. Two of the park's species of amphibians may now be extirpated from the park; one explanation is pesticide poisoning. The California red-legged frog and the foothill yellow-legged frog were historically present in the park but now may be gone.

Another key to improving naturalness is removing exotics. The park is now home to several exotic plants including spotted knapweed and bull thistle, which threaten native flora and fauna. Non-native bullfrogs and introduced fish also affect native wildlife. In fact, of the twelve fish species listed as present in the park, six were introduced. Two native species may already be extirpated.

ZION
Sculptured Canyons and Soaring Cliffs

Springdale, Utah

ACRES: 146,598

Zion Canyon was home to prehistoric hunter-gatherers as early as 6000 B.C.; cultivation of the canyon began about 300 B.C. One wonders what these first inhabitants of Zion Canyon thought of their homes. Did they perceive the scenic beauty of the two thousand-foot-deep canyon? Were the colorful canyon walls spiritually uplifting or merely an impediment to travel?

More than two thousand years later people were still trying to make a living in Zion Canyon. In 1847 the Mormons moved into Utah, and soon pioneers were sent to southern Utah to settle the territory and grow cotton in Utah's "Dixie." One of the settlements, Springdale, was located just outside Zion Canyon. In 1863, Isaac Behunin built the first cabin in the canyon, soon to be followed by others. For the remainder of the nineteenth century, Mormon settlers eked out a living in the canyon, facing floods and poor soils.

In the early twentieth century, southern Utah was becoming recognized as a tourist attraction. In

1909, Zion Canyon was designated as Mukuntuweap National Monument. Getting to it was another matter, as roads into the canyon were mostly impassable. Over the next decade, road and rail access was improved, and in 1919 Congress established Zion National Park. The Kolobs Canyon area was made a national monument in 1937 and added to Zion National Park in 1956. Today, the park contains 146,598 acres; 90 percent of the park has been proposed for wilderness designation.

Zion contains several plant communities. The southern part of the park is desert characterized by colorful mesas and canyons. The park's low point is Coalpits Wash, 3,666 feet in elevation. Ponderosa pine and aspen cover the Kolobs Canyon area in the northwestern part of the park; elevations here are much higher and include the park's high point, 8,726-foot Horse Ranch Mountain. Riparian habitats are found in the park, including the Virgin River, the creator of Zion Canyon.

There are seventy-eight species of mammals in the park, the most numerous of which are the rodents. Among the squirrels, the most common are the white-tailed antelope squirrel and the rock squirrel. Tree squirrels include the red squirrel and the northern flying squirrel. Zion is home to three chipmunk species: the Uinta chipmunk, least chipmunk, and cliff chipmunk.

Beaver are common in some of the park's rivers and streams. Two species of woodrat, or pack rat, inhabit the park, the desert woodrat and the bushy-tailed woodrat. Wherever there are trees, a porcupine may be commonly found.

Among the lagomorphs, the desert cottontail is most common. Others include the mountain (or Nuttall's) cottontail, the black-tailed jackrabbit, and the American pika.

The park's largest predator is the black bear. It migrates through the high country and is rare in Zion. The mountain lion is fairly common throughout the park but few people see it. The bobcat is nocturnal and seldom seen.

The coyote is common but more often heard than seen. The gray fox is nocturnal and seen only occasionally along park roads. Both the kit fox and red fox are rare. The most common of the smaller carnivores is the ringtail. The striped skunk is completely nocturnal and seldom seen.

The mule deer is the most common of the ungulates and occurs throughout the park. Elk are uncommon residents of the higher elevations. The original population of desert bighorn sheep was extirpated, but in 1973 twelve were returned to the park and placed in an enclosure. They were later re-

leased in the valley of the East Fork of the Virgin River, and their descendants remain in steep areas in the eastern portion of the park.

CONSERVATION CONCERNS

During 2003, 2,451,977 people visited Zion National Park, and visitation is likely to increase. Zion is not far from Las Vegas and closer to Los Angeles than Yosemite National Park; the park is not as remote as it may seem. On busy days in July and August, eleven thousand people enter the park daily.

Prior to 2000, visitors would attempt to drive up Zion Canyon only to find parking lots full and traffic clogged. In May 2000, a mandatory shuttle system was initiated that prohibits private vehicles in the upper six miles of Zion Canyon, now called the Zion Canyon Scenic Drive. These restrictions are permanent annually from April through October. Shuttles stop at the visitor center, campgrounds, and in Springdale with parking at each stop. In 2003, 2,417,000 people rode the shuttle, an increase of 2.6 percent from the previous year.

Air quality remains high in the park. No serious air pollution exists to date. Visibility has not been impaired as of yet. Park staff is monitoring various air quality parameters and participating in regional planning efforts where air quality is an issue. Similarly, park waters are of high quality and are free flowing. Measurements are taken to ensure that water quality remains high.

Water, canyons, and sharp peaks are everywhere in Zion. Here, the North Fork of the Virgin River flows over one of the gentler slopes.

The Mammals

MARSUPIALS

Marsupials are commonly called "pouched mammals," but not all marsupials have a pouch (called a *marsupium*). Marsupials are best distinguished from placental mammals based on the very short developmental period of the young prior to birth—only eight to forty days. All marsupials are so small and undeveloped at birth that, regardless of the species, the entire litter weighs less than 1 percent of the mother's body weight. In placental mammals such as shrews and mice a litter may equal 50 percent of the mother's weight. Most of the early growth and development of marsupials does not take place in the uterus but instead occurs during nursing (lactation), while the young are firmly attached to a nipple in the pouch. Marsupials also differ from placental mammals of equivalent size in having lower metabolic rates, slower growth, and many skeletal and anatomical differences including smaller brains. Although we often use the commonly accepted term *placental mammals* to describe nonmarsupials, it should be discouraged because it implies that marsupials do not have a placenta. They have a placenta, but it is structurally simpler and less efficient than that found in placental mammals (technically called *eutherians*). Marsupials (technically called *metatherians*) represent a group that diverged from eutherian mammals about 130 million years ago. Since then, marsupials, including the Virginia opossum, have been quite successful in adapting to a variety of habitats and environmental conditions.

Virginia Opossum

Although most people think of marsupials as occurring primarily in Australia, 25 percent of the approximately 270 recognized species are in Central and South America. The Virginia opossum is the only species of marsupial in the United States and the national parks. With its white face, narrow, pointed snout, leathery ears, and naked, scaly prehensile (grasping) tail, no other mammal in the national parks is quite like an opossum. If seen quickly late at night, when the opossum is most active, it could easily be mistaken for a large rodent. It is about the size of a domestic cat, with head and body length equal to that of the tail. The tail is black near the body and white toward the end. The hind feet of the opossum have opposable big toes. Like the tail, they are useful for grasping. The opossum has

fifty teeth, more than any other species of terrestrial mammal in the United States. The canines of males are formidable and noticeably longer than those of females. Body coloration is generally grayish with long black guard hairs. Different color variations occur geographically, including all-white or black individuals. The species occurs from Costa Rica north through Mexico, throughout much of the eastern United States to southern Ontario, and in the western United States to southern British Columbia. Opossums have increased their range northward during the past hundred years but are limited by snow and freezing temperatures. Individuals in the northern parts of the range often have ears and tails damaged by frostbite.

Virginia opossums move about easily in trees but spend most of their time on the ground. They prefer deciduous forest habitat close to water, where they den in hollow trees, under logs, in rocky outcrops and crevices, culverts, or old buildings. They also may occupy abandoned burrows constructed by armadillos, skunks, woodchucks, or other species. As many homeowners can attest, opossums also do quite well in suburban areas. They are not social animals and do not form groups. Individuals rarely move more than a half mile from their dens as they forage, although they may change den sites depending on food availability. Opossums do not hibernate, but during cold weather they may not leave the den for several days. Park visitors are most likely to see them as they are foraging at night. They eat fruits, berries, and vegetables, as well as insects, small mammals, birds, eggs, snakes, and carrion. Interestingly, opossums seem to have a remarkable resistance to the venom of poisonous snakes on which they prey. Their omnivorous habits often lead them to campsites in parks, where people may encounter them as late-night visitors to trash cans.

If approached, opossums may engage in threat display. They will hiss but rarely bite. When threatened or alarmed, they are best known for "playing possum"—their tendency to feign death. An opossum will enter a temporary catatonic condition, fall limp on its side, open its mouth, and loll its tongue while drooling copious amounts of saliva. The animal may also defecate and discharge a vile-smelling

VIRGINIA OPOSSUM	
Didelphis virginiana	
COMMON NAMES	Opossum, Possum
HEAD AND BODY LENGTH	12 inches (30 cm)
TAIL LENGTH	12 inches (30 cm)
BODY WEIGHT	2–11 pounds (1–5 kg) males
	2–8 pounds (1–4 kg) females

As this Virginia opossum moves among tree branches, its prehensile tail functions as a fifth limb. The only marsupial in the United States, the opossum occurs in national parks from Florida to California.

(Pages 98–99) Grizzly bears feed on different animals and plants as the seasons change; for example, they shift to a salmon diet when the spawning season begins. Although hunted in Alaska and Canada as a game animal, grizzlies south of Canada are a protected species.

greenish substance from the anal glands. If left alone for a few minutes, however, the opossum quickly "recovers" and runs away. This behavior probably helps deter predators largely because of the odorous discharge from the glands. Nonetheless, a variety of terrestrial carnivores, especially coyotes, foxes, bobcats, and dogs, as well as owls, prey on opossums.

The breeding season begins around January and can continue through much of the year, especially in southern regions. Females breed when less than a year of age. Gestation in all marsupials is very short, but in the opossum is remarkably so—only 12.5 days. As such, the newborn (neonate) is very small and undeveloped. It is about one-half inch long, and weighs only 0.005 ounce (it would take 3,200 newborns to equal one pound). After birth, the forelimbs are used to grasp the mother's belly fur because the neonate must pull itself up into the pouch. Once there, it attaches to a nipple that swells in its mouth and helps hold it in place. Females usually have thirteen nipples. A neonate that fails to attach to one—either because it did not make it into the pouch or because litter size exceeded thirteen and no nipple was available—will die. However, average litter size in most areas is seven or eight. Young nurse in the pouch for about two months. They are capable of eating solid food after this time, but weaning usually does not occur until they are about three months old. Once out of the pouch, young may ride on their mother's back, but as soon as they are independent they disperse. Most individuals in the wild live less than two years.

CONSERVATION CONCERNS

As the sole marsupial representative in North America, opossums are a fascinating addition to the mammalian fauna. They are often very common locally and face no serious threats. They occasionally prey on domestic poultry and ground-nesting waterfowl, but generally are of little significance to wildlife managers or conservation programs.

SHREWS AND MOLES

Shrews and moles are intriguing members of the mammalian fauna. Common and widespread in most national parks, they will rarely if ever be seen by visitors. Close to 10 percent of the 312 species of shrews worldwide occur in North America, from Alaska through Mexico. Several species, such as cinereus shrews and montane shrews, have extensive geographic ranges and occur within many national parks. Others are very restricted geographically, including Preble's, Pacific, and Inyo shrews (see Table, pages 200–201). Generally solitary, shrews occupy forest, grassland, marsh, or desert habitats but are almost never observed as they forage for insects in leaf litter and dense understory. They have tiny eyes, long, pointed snouts, and short, dense dark fur. Shrews range in mean body weight from about 0.1 ounce (the weight of a dime) for pygmy shrews up to 0.9 ounce for northern short-tailed shrews. Because they are so small, shrews have extremely high metabolic rates. The normal heart rate of pygmy shrews, for example, is 1,030 beats per minute. As a result, they must forage throughout much of the day and night.

Shrews are too small to hibernate or migrate, and their average life span is only about one year. They

Among the smallest mammals in the world, shrews are rarely seen as they forage beneath leaf litter in their constant search for insect prey.

are fascinating animals not only because of their tiny size but also because of many interesting adaptations. Short-tailed shrews secrete a toxin from one of their salivary glands that they use to immobilize invertebrate prey. Shrews also use echolocation. They emit high frequency sound pulses and, from the returning echoes, maneuver through habitat, communicate with other shrews, and detect prey. As noted, park visitors will rarely see shrews, and identification is difficult—experts often base identification of different species on the size and shape of their teeth. Shrews remain among the least known and most poorly documented mammals in the national parks.

Moles are related to shrews and are in the same order (Insectivora). Worldwide, there are forty-two species of moles, but only seven of these occur in the United States. Unlike shrews, moles spend almost their entire lives burrowing underground (they are fossorial). As you might expect, they have some dramatic adaptations for life underground including tiny eyes and no external ears. Their velvety fur moves equally well in any direction, making it easy for them to move forward or backward in a tight tunnel. Moles also have very short, extremely powerful forelimbs with broad, flat forefeet for digging. Because they spend so much time underground, sight and hearing are reduced, but their senses of touch and smell are highly developed.

The unique star-nosed mole, which occurs in several national parks (see Table, pages 200–201), is a tactile specialist. It has a ring of twenty-two fleshy, highly sensitive tentacles surrounding the nose that are used to locate prey items.

The only evidence most visitors will see of moles in the national parks or elsewhere is the surface ridges that appear from their tunneling activity and resulting molehills about seven inches high. Tunnels are a little less than two inches in diameter and form networks that can extend throughout an area the size of a football field. Molehills differ from the mounds of pocket gophers in being cone-shaped, with the cone directly over the tunnel opening, and having finer-grained dirt.

CONSERVATION CONCERNS

Shrews and moles are useful because of the number of invertebrates they consume and because they aerate and loosen soil. As many homeowners are well aware, moles also damage yards and gardens. Aside from the fact that a few species of shrews are very limited in geographic distribution and are considered endangered, these interesting additions to the mammalian fauna of the national parks are of no significance to wildlife managers in conservation programs.

BATS

Although people speak of "flying lemurs" and "flying squirrels," these are misnomers—the only mammals that fly are bats. There are about 920 species of bats worldwide, and they account for about 20 percent of all living mammalian species in the world today. Most aspects of the natural history and physical structure of bats relate to flight. Their forearms and fingers are elongated and covered with skin to form a wing. The name of the order, Chiroptera, means "hand wing." Bats usually have an additional membrane between their legs, called an interfemoral membrane or uropatagium. This membrane encloses the tail and provides additional surface area for flight. Compared to birds, bats are slow fliers but highly maneuverable. Flight has allowed bats to specialize in feeding, reproduction, and behavior. Some bats prey on fish or other small vertebrates such as frogs or mice, others take only fruit, and a few species of vampire bats consume only blood. Most bats in the United States and the national parks are insectivorous, taking small moths and insects in flight. A few species consume nectar and pollen from flowers. Interestingly, the shape of a species' wing is closely related to its feeding habits. Certain species forage in areas with dense, obstructing vegetation where slow flight and high maneuverability are at a premium. These species have wings that are wide relative to their length. Conversely, relatively long, thin wings allow for higher, faster flight.

Regardless of their feeding niche, all bats in North America use echolocation to maneuver through habitat (including the inky blackness of deep caves and mines), avoid potential predators, and find prey. They emit very high frequency sound pulses and gain information about their surroundings based on characteristics of the returning echoes. Bat echolocation is a highly sophisticated system. Echolocation gives a bat the same amount of information about size, shape, texture, distance, and movement of objects in its immediate environment as vision does—everything except color. Bats also have completely functional vision; the term "blind as a bat" is simply not true. However, their primary means of perception is through the auditory orientation afforded by echolocation, an excellent adaptation for animals actively foraging at night. When resting during the day, bats roost in a variety of protected sites. Species such as big brown bats, little brown bats, and pipistrelles roost in buildings, caves, and abandoned mines. The endangered Indiana bat

(Top) One of the most impressive sights at Carlsbad Caverns National Park is the evening emergence of hundreds of thousands of Brazilian free-tailed bats as they leave their roost in search of insect prey. Colonies of these bats form the largest aggregations of any vertebrate species in the world.
(Right) The insectivorous species of bats use their highly sophisticated echolocation system and large ears to maneuver over the landscape in search of prey.

roosts under the loose bark of trees. Red bats, hoary bats, and silver-haired bats roost under tree limbs or vegetation. Certain species roost by themselves, but others form large colonies, such as the Brazilian free-tailed bats found in Carlsbad Caverns National Park. Temperature is a critical aspect in where bats choose to roost; diet, social structure, and risk of predation are other factors.

Bats also demonstrate intriguing reproductive adaptations. Most North American bats mate in the fall, prior to hibernation or migration, and females store viable sperm in the uterus throughout the winter. This reproductive mode is called delayed fertilization because ovulation and subsequent fertilization occur in late winter or early spring. Young are then born at about the same time as the emergence of insect prey. Most bat species have one young per year. This is unusual in small mammals, which generally have large numbers of young per litter and several litters per year. Young bats (pups) are born in a maternity colony consisting only of adult females. Bats also are unusual in that individuals have much longer life spans—approaching twenty-five to thirty years—than is typical for small mammals.

Bats in the national parks are primarily nocturnal; visitors often may see them foraging around a light at night, dusk, or dawn. Nonetheless, visual identification of most species as they forage is practically impossible, even for experts. For that reason, we do not discuss individual species except the Brazilian free-tailed bat, which draws hundreds of thousands of visitors annually to Carlsbad Caverns National Park. Bat species known or presumed to occur in each national park are listed in the Table (pages 201–2).

Brazilian Free-tailed Bat

One of the more spectacular sights in any of the national parks is the evening emergence at Carlsbad Caverns National Park of Brazilian free-tailed bats, by far the most numerous of the sixteen species of bats known to occur there. From a distance, it appears as if a huge plume of smoke is rising into the sky. In reality, this plume is 250,000 to 300,000 bats leaving their roost in the caverns as they begin their evening foraging activities. The emergence of Brazilian free-tailed bats from Carlsbad Caverns is greatest in August and September when the young are able to fly. But emergence is not the only time to witness this biological phenomenon. Prior to dawn, visitors can watch as bats return to the caverns from their foraging trip, swooping down from great heights at speeds of 25 miles per hour or more. Brazilian free-tailed bats also occur in several other national parks (see Table) and are one of the most abundant, widely distributed bat species in North America. They range across the southern half of the United States from coast to coast, throughout Mexico and Central America, and in most Caribbean islands.

Brazilian free-tailed bats have dark gray-brown fur and black wings. Wingspan is about twelve inches and makes them appear larger than they really are. The wings are relatively long and narrow, allowing for fast, high flight over long distances. In fact, individuals may travel thirty miles or more between roost and foraging sites, and some bats migrate eight hundred miles to Mexico or Central America for the winter. The species has been documented to reach altitudes as high as ten thousand feet. Like other free-tailed bats (family Molossidae), the last half of the tail in Brazilian free-tailed bats extends beyond the uropatagium—the membrane between the hind legs—and is "free." In most species of bats in the United States (family Vespertilionidae), the tail is completely enclosed within the uropatagium.

Moths and flying insects are the major prey items, and an individual bat consumes up to 50 percent of its body weight foraging each night. Thus, the colony in Carlsbad Caverns probably consumes close to nine thousand pounds of prey a night, including many agricultural pests. Brazilian free-tailed bats give birth to one young each year in June following a gestation of ninety to one hundred days. Pups are left together in a large, dense cluster (called a *creche*) of five hundred or more individuals per square foot on a cave wall. Mothers roost in separate maternity colonies but visit their pups several times a day to nurse. Mothers apparently find their own young among the mass of thousands of pups through sound communication, scent, and by remembering their general location within a cluster. Young are able to fly when they are four to five weeks of age.

CONSERVATION CONCERNS

Brazilian free-tailed bats have been roosting in Carlsbad Caverns for more than five thousand years. During the 1930s, it was estimated that there were eight to nine million individuals there. Today, that number has declined precipitously to approximately 350,000, a figure that fluctuates annually. Other tremendously large aggregations of this species are known, including Bracken Cave, Texas, with a colony of twenty million bats. At one time, Eagle Creek Cave in Arizona may have had a colony as large as fifty million individuals. These are the largest aggregations of any mammalian (or any vertebrate) species in the world. Such large concentrations of individuals in relatively few places make the species particularly vulnerable to major disturbances. Reductions in these tremendous densities in recent decades are believed to be related to loss of foraging habitats, destruction of roost sites, and accumulation of pesticides, especially DDT. Although this chemical was banned in the United States in 1972, it is still used in Mexico, where the bats spend the winter. Many other species of bats have declined dramatically in density during the last forty years because of the same human-induced problems. As a result, several North American species are on the federal Endangered Species List, including the Indiana bat and the gray bat.

BRAZILIAN FREE-TAILED BAT *Tadarida brasiliensis*	
COMMON NAMES	Mexican free-tailed bat, Free-tailed bat, Guano bat
HEAD AND BODY LENGTH	3 inches (7.6 cm)
TAIL LENGTH	1.3 inches (3.3 cm)
BODY WEIGHT	0.3–0.5 ounces (8.5–14 g)

NINE-BANDED ARMADILLO

The nine-banded armadillo is very distinctive—it cannot be confused with any other North American mammal. Spanish conquistadors named it "little armored one," which refers to the familiar gray-brown to yellow-tan shell (carapace) covering the upper part of the body. This carapace is made up of scales called scutes formed from bone and connective tissue. A solid shield covers the head, shoulders, and rump. The back has a series of overlapping bands (usually nine but sometimes seven or eight) connected by skin. The legs and tail also are protected by armor. Only the underside of an armadillo remains unprotected. This is where the hair is most dense, although a few hairs also protrude from the edges of individual scutes. The armadillo has short powerful legs with long heavy claws used for digging. Especially in soft soil, it can quickly escape danger by excavating rapidly. It has a long tapered snout and no incisor or canine teeth but only simple peg-like cheek-teeth. The nine-banded armadillo is most active at night, when it feeds on insects, vegetation, bird eggs, and small vertebrates.

The nine-banded armadillo is the sole North American representative of the order Xenarthra. All xenarthrans are restricted to the New World. In addition to several other species of armadillos throughout Central and South America, this order includes two-toed and three-toed sloths and various species of anteaters. This sounds like an odd assortment of mammalian species with little in common. But they are related because all of them have structurally complex vertebrae. The order was formerly called Edentata, meaning "without teeth," but that is a misnomer because only the anteaters have no teeth.

The nine-banded armadillo has expanded its range significantly north and east in the United States during the last two hundred years. It now occurs in much of the southeastern United States, throughout Texas, the Gulf Coast, and through introductions to Florida. In the Midwest, it is found as far north as Missouri, with a few individuals in southern Illinois. Like other xenarthrans, the nine-banded armadillo is sensitive to cold temperatures, and this probably limits the northward expansion of its geographic distribution.

Reproduction in armadillos is especially intriguing. They undergo delayed implantation, meaning the fertilized egg (zygote) floats free in the uterus prior to implanting. This delay is typically for three to four months, although in exceptional cases up to a two-year delay has been documented. Nine-banded armadillos also are remarkable because litters generally are quadruplets and all the same sex. This occurs because the fertilized egg splits apart at the four-cell stage and gives rise to four separate and identical embryos. This phenomenon (technically called *monozygotic polyembryony*) is known to occur only in armadillos. Newborn young each weigh about four ounces and have a very thin, pliable carapace. They are precocial, with eyes open, and are able to walk very quickly. Young develop rapidly and are on their own within three to four months.

CONSERVATION CONCERNS

Armadillos exhibit both positives and negatives as a wildlife resource. In some areas they seriously damage agricultural crops as well as lawns and gardens. They also prey on the eggs of ground-nesting birds including quail and turkeys, and in Florida may have a negative impact on endangered species through foraging and predation. Conversely, armadillos are valuable animals for medical research because of their reproductive physiology. They are also important because they are the only species other than humans that naturally gets leprosy. Throughout their range, armadillos are commonly killed on roads because of their habit of jumping straight up when alarmed. This defensive reaction is useful when threatened by predators such as coyotes, but it is unfortunate when they are startled by an oncoming vehicle and jump about bumper height.

Related to sloths and South American anteaters, the nine-banded armadillo is a unique mammal found throughout parts of the southern United States. It occurs in both Hot Springs and Everglades National Parks.

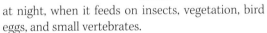

NINE-BANDED ARMADILLO *Dasypus novemcinctus*	
COMMON NAMES	Armadillo, Long-nosed armadillo
HEAD AND BODY LENGTH	16 inches (40.6 cm)
TAIL LENGTH	12 inches (30 cm)
BODY WEIGHT MALES	14 pounds (6.4 kg)
BODY WEIGHT FEMALES	10 pounds (4.5 kg)

LAGOMORPHS

Growing up with stories of Peter Rabbit and cartoons of Bugs Bunny, most people certainly recognize rabbits and hares. Along with pikas, they comprise the mammalian order Lagomorpha, which means hare-shaped. Most of the eighteen species of lagomorphs in North America are easily seen or heard in national parks. Less obvious, however, are the physical and behavioral differences that separate rabbits and hares. Differences include skull structure and the number of chromosomes (rabbits generally have forty-two, and hares usually have forty-eight). Hares generally have larger hind feet relative to their body size than rabbits do. Also, newborn rabbits are naked, without fur, and helpless (altricial), whereas newborn hares are fully furred, with eyes open, and are capable of running within hours after birth (precocial). Rabbits give birth in well-constructed, often fur-lined nests; hares give birth in shallow depressions on the ground, called forms. Park visitors, however, will not see the unique characteristic that distinguishes lagomorphs from all other mammals. They have two pairs of upper incisors—a large, prominent pair of teeth directly in front of a smaller peg-like pair. Interestingly, pikas, rabbits, and hares all reingest their fecal pellets, a practice called coprophagy, which allows them to assimilate nutrients that would otherwise be wasted.

Pikas

Only two species of pikas occur in North America. Collared pikas are distributed throughout south central Alaska and northwestern Canada. American pikas occur in the western United States, British Columbia, and Alberta. Both species inhabit high elevation rocky talus slopes and meadows, often above timberline. Unlike rabbits and hares, pikas can easily be mistaken for large rodents. With their short legs, small rounded ears, and inconspicuous tails, pikas closely resemble guinea pigs. Their grayish-brown fur is soft and dense. In the collared pika, the "collar" is formed by an indistinct band of grayish fur around the top of the neck.

Pikas have some intriguing behaviors. They are active throughout the day, and spend much of that time cutting and carrying a variety of vegetation back to their individual territories. The vegetation is dried and stacked, forming "haypiles" that may reach two feet high. These caches of stored vegetation are used by individual pikas throughout the harsh winter. Pikas do not hibernate, and additional foraging often occurs as they burrow under snow to reach plants. Although pikas in the national parks may be seen scurrying over the rocks, they more often will be heard. High-pitched short and long whistles are given by both sexes. Whistles communicate alarms between individuals when predators are sighted and help to maintain territorial boundaries. Pikas also demarcate territories through scent marking by rubbing their cheek glands on rocks.

Both species of pikas produce two litters per year. Average litter size is about three altricial young following a gestation of thirty days. Once they are independent at about a month of age, it is essential that juveniles find a vacant territory and avoid aggressive interactions with resident adults. Pikas are relatively safe in rocky talus slopes but nonetheless are susceptible to predation by coyotes, foxes,

Although in appearance it resembles a rodent, the pika is closely related to rabbits and hares. This individual is collecting vegetation to store for the upcoming winter.

COLLARED PIKA	
Ochotona collaris	
COMMON NAMES	Cony, Rock rabbit, Whistling hare
HEAD AND BODY LENGTH	7–8 inches (18–20 cm)
TAIL LENGTH	Inconspicuous
BODY WEIGHT	4–5 ounces (113–142 g)

AMERICAN PIKA	
Ochotona princeps	
COMMON NAMES	Southern pika, Rock rabbit, Piping hare, Mouse hare, Haymaker
HEAD AND BODY LENGTH	6–8 inches (15–20 cm)
TAIL LENGTH	Inconspicuous
BODY WEIGHT	4–6 ounces (113–170 g)

Eastern Cottontail
Sylvilagus floridanus

COMMON NAMES	Florida cottontail
HEAD AND BODY LENGTH	15 inches (38 cm)
TAIL LENGTH	2 inches (5 cm)
BODY WEIGHT	1.7–3.3 pounds (0.8–1.5 kg)

Appalachian Cottontail
Sylvilagus obscurus

HEAD AND BODY LENGTH	14 inches (36 cm)
TAIL LENGTH	1.8 inches (4.5 cm)
BODY WEIGHT	1.6–2.2 pounds (0.7–1.0 kg)

Mountain Cottontail
Sylvilagus nuttallii

COMMON NAMES	Nuttall's cottontail
HEAD AND BODY LENGTH	12 inches (31 cm)
TAIL LENGTH	2 inches (5 cm)
BODY WEIGHT	1.3–2.0 pounds (0.6–0.9 kg)

Desert Cottontail
Sylvilagus audubonii

COMMON NAMES	Audubon's cottontail
HEAD AND BODY LENGTH	13 inches (33 cm)
TAIL LENGTH	2 inches (5 cm)
BODY WEIGHT	1.6–2.9 pounds (0.7–1.3 kg)

Pygmy Rabbit
Brachylagus idahoensis

HEAD AND BODY LENGTH	10 inches (25 cm)
TAIL LENGTH	0.8 inch (2 cm)
BODY WEIGHT	0.9–1.0 pound (409–454 g)

Brush Rabbit
Sylvilagus bachmani

COMMON NAMES	Bachman's cottontail
HEAD AND BODY LENGTH	12 inches (31 cm)
TAIL LENGTH	1 inch (2.5 cm)
BODY WEIGHT	1.1–2.0 pounds (0.5–0.9 kg)

Marsh Rabbit
Sylvilagus palustris

HEAD AND BODY LENGTH	16 inches (41 cm)
TAIL LENGTH	1.4 inches (3.6 cm)
BODY WEIGHT	2.6–4.8 pounds (1.2–2.2 kg)

Swamp Rabbit
Sylvilagus aquaticus

COMMON NAMES	Canecutter
HEAD AND BODY LENGTH	17 inches (43 cm)
TAIL LENGTH	2.4 inches (6 cm)
BODY WEIGHT	3.5–6.0 pounds (1.6–2.7 kg)

martens, and raptors. The primary predators, however, are small weasels that are capable of following pikas through cracks and crevices among the rocks.

CONSERVATION CONCERNS
Pikas are tiny, charismatic, delightful mammals but are of no economic importance or significance in terms of management or conservation initiatives.

Rabbits

Of the nine species of rabbits in the United States, eight occur in national parks. Many of these species are commonly referred to as "cottontails," but all rabbits share very similar characteristics familiar to most people. These include long ears and hind legs, large eyes, a short tail, somewhat nervous behavior, and rapid, erratic hopping toward the nearest cover when startled or threatened.

The best known and most widespread species is the eastern cottontail—a common resident of suburban parks and neighborhoods. It occurs throughout the eastern half of the United States (except in most of New England) and has been introduced into parts of the southwest and northwest as well. Eastern cottontails are very adaptable, and there are few habitat types in which they cannot be found, although old fields are preferred. Coloration varies somewhat with locality but fur is grayish to brownish, often with dark hairs giving a penciled effect, grading to lighter coloration on the belly. Their geographic range overlaps that of seven other species of rabbits. In the eastern United States, eastern cottontails are very difficult to distinguish from New England cottontails and recently described Appalachian cottontails. As the name implies, New England cottontails are restricted to parts of southern Maine and south through eastern New York, often in higher elevation forested areas. Their range apparently does not extend far enough north for them to occur in Acadia National Park, however. Until recently, the Appalachian cottontail was considered to be the same species as the New England cottontail. The geographic range of the Appalachian cottontail extends discontinuously throughout the Appalachian Mountains in higher elevation forested and meadow areas from Pennsylvania south to Alabama, and does not overlap that of the New England cottontail. Appalachian and New England cottontails can be differentiated only on the basis of skull or genetic characteristics; likewise, as far as size and pelage (fur), both species look the same as eastern cottontails.

Several different species of rabbits occur throughout the United States. Differing in subtle ways, they are usually referred to simply as "cottontails." They are among the most easily seen and recognized mammals in the national parks.

Two other species of rabbits are referred to as cottontails. Mountain cottontails also look similar to eastern cottontails, but their ranges do not overlap. Mountain cottontails occur throughout the intermountain west from southern Canada to Arizona and New Mexico, between the Rocky and Cascade mountain ranges. They occupy rocky sagebrush habitats, as well as wooded areas, usually associated with water sources. Like all rabbits, females are slightly larger than males. Desert cottontails are found in predominantly arid areas of the intermountain west and southwest United States, from the lowest elevations of Death Valley National Park to high elevation wooded mountain habitats. The ears of desert cottontails are slightly longer than those of either mountain or eastern cottontails, and act as radiators to effectively dissipate body heat.

Although they are not referred to as cottontails, the other species of rabbit found in national parks is very similar in habits and appearance. The brush rabbit occurs from the Columbia River, south along the Pacific Coast, and west of the Cascades to the southern tip of Baja California. It is associated with dense cover and uses burrows constructed by other species. The brush rabbit is fairly small, with relatively short legs and ears. Very secretive, it rarely travels far from cover and is less readily seen than some of the other species.

Swamp rabbits and marsh rabbits are restricted to the southeastern United States and, given their common names, it is no surprise that both species are good swimmers and are associated with water. Marsh rabbits occur in low elevation fresh and brackish wetland areas from Virginia to southern Florida and west to southern Alabama. Unlike cottontails, marsh rabbits have reddish belly fur. The species is smaller than the swamp rabbit but heavier than the cottontails. Swamp rabbits extend throughout much of the Gulf Coast area to east Texas and north to southern Missouri, Illinois, and Indiana, but do not overlap the range of marsh rabbits. Swamp rabbits are closely associated with wetlands and are never far from water, which they readily enter to escape potential predators. Swamp rabbits are large and chunky, with reddish-brown to black pelage.

Pygmy rabbits are the smallest of all. They once occurred throughout the Great Basin region with their range overlapping those of several other species of rabbits and hares. Loss of sagebrush habitat on which pygmy rabbits depend has significantly reduced their distribution. Pygmy rabbits are unique not only because of their small size, but unlike other rabbits they burrow extensively and give a definite alarm call.

Rabbits are generally active at dawn and dusk but also throughout the day as they consume vegetation. Geographic area, elevation, habitat types, and season of the year all determine what a species eats. Grasses and broadleaf herbaceous plants

usually are a large part of any diet. Rabbits remain concealed in vegetation, brushpiles, or rocks and are cautious as they venture out to feed. Population density of rabbits is quite variable, and generally follows a long-term nine- to eleven-year cycle. As rabbit populations increase during this period, they reach very high densities before they "crash," with almost no individuals in an area. Population densities of predatory species that rely primarily on rabbits follow a year or two behind this cyclic pattern.

Reproductive patterns are similar in all rabbits. All have short gestation (four to five weeks), large litter sizes (generally three to four but up to eight in some species), and numerous litters per year (usually three but up to six or seven in certain species in some areas). In general, rabbits in northern areas have larger litter sizes, but fewer litters per year, than individuals in the south. It is easy to understand the expression "breed like rabbits." A female eastern cottontail in the southern portion of the range may breed throughout the year and produce thirty-five young. Newborn cottontails are about four inches long and weigh a little more than an ounce. Young are born in rounded, cup-shaped nests lined with dry grass and leaves, as well as fur the mother pulls from her belly. A plug of fur or vegetation conceals the young while they are in the nest, and the mother returns periodically to nurse them. Young leave the nest two to three weeks after birth. The average lifespan of rabbits is less than a year. Numerous species prey on them including coyotes, foxes, bobcats, weasels, hawks, and owls.

CONSERVATION CONCERNS

In terms of the number of people who hunt them and the number of animals taken each year, eastern cottontail rabbits are the number-one game mammal in the United States. Other species of rabbits also provide recreational hunting opportunities, as well as a major prey base for numerous species of mammalian and avian predators. Habitat management by state or federal agencies to maintain or increase population densities of rabbits generally involves retarding vegetation succession to provide adequate forage and escape cover. Each state sets regulations on the length of the hunting season and bag limits. When population densities are too high, rabbits may significantly damage agricultural crops, as well as suburban gardens and landscaping. As noted, pygmy rabbit populations have declined with loss of sagebrush habitat because of livestock grazing. The species is now endangered, as is the New England cottontail, which has been displaced in many parts of its former range by the eastern cottontail.

Hares

Five species of hares—often called jackrabbits—occur in national parks. "Jackrabbit" or "jackass rabbit" refers to the very long ears most hares have, similar to those of a jackass. This is an unfortunate and confusing common name because they are not really rabbits.

Snowshoe hares enjoy the widest geographic distribution of any North American hare. They occur in coniferous forest and dense brushy habitats throughout most of Alaska, mainland Canada, the New England states south through the Appalachian Mountains, northern tier states, and much of the western United States. During the summer, the fur is reddish brown on the back and head and grayish on the belly, and the ears have black tips. Following the autumn molt, the fur is completely white except for the black-tipped ears—an adaptation for remaining less visible in snowy winter conditions. At lower elevations or in more southerly regions without heavy snow (such as Olympic National Park), winter pelage does not turn white. The common name refers to their disproportionately large, well-furred hind feet, which enable snowshoe hares to cross deep, soft snow in the winter. Active throughout the year, females produce two to five litters annually. Average litter size ranges from two or three in southern parts of the range to five or six in the north. Gestation is five to six weeks, and newborns can run soon after birth. Weaning occurs at about one month of age. As in rabbits, ten-year population cycles occur throughout the geographic range. Snowshoe hares increase in density for several years followed by a rapid population "crash." Although no one knows for certain what causes these cycles, it may be an interaction of food resources and predation. Numerous predators take snowshoe hares, including foxes, coyotes, bobcats, and various raptors. The population densities of lynx, which rely heavily on snowshoe hares for food, follow a cycle about a year behind that of the hares.

Like the snowshoe hare, the Alaskan hare molts from a brown pelage in the summer to an all-white winter pelage. It occurs from northern Alaska throughout the western coast, the Seward Peninsula, and most of the Alaskan Peninsula. The species inhabits tundra and coastal lowlands, and uses brushy thickets for cover. Because of the short summer season, a single litter of about six young is typical after the snows melt. To increase their chances of surviving the coming winter, young grow very quickly, aided by prolonged nursing of up to nine weeks. The Alaskan hare is vulnerable to predation by a variety of avian and mammalian species.

Black-tailed jackrabbits are widely distributed throughout the western United States from Washington to South Dakota and south through Texas, and in California, into Baja California and Mexico. The species has also been successfully introduced into several eastern states. Blacktails are quite adaptable and occupy a variety of habitats from sea level to high elevation, although they are typically found in arid rangelands or agricultural areas. They consume grasses, shrubs, and a variety of plants, and can subsist on poor-quality forage that is unpalatable to other species. Black-tailed jackrabbits usually forage at night, spending the hotter portions of the day concealed in the shade

A snowshoe hare, in white winter pelage, reaches high to feed on a pussy willow twig. Snowshoe hares are widely distributed in North America and occur in about twenty-five national parks.

BLACK-TAILED JACKRABBIT *Lepus californicus*	
COMMON NAMES	California jackrabbit
HEAD AND BODY LENGTH	16–20 inches (41–51 cm)
TAIL LENGTH	2–4 inches (5–10 cm)
BODY WEIGHT	3–7 pounds (1.4–3.2 kg)

WHITE-TAILED JACKRABBIT *Lepus townsendii*	
COMMON NAMES	Prairie hare
HEAD AND BODY LENGTH	19–21 inches (48–53 cm)
TAIL LENGTH	3–5 inches (8–13 cm)
BODY WEIGHT	5.5–9.5 pounds (2.5–4.3 kg)

ANTELOPE JACKRABBIT *Lepus alleni*	
COMMON NAMES	Allen's hare, Saddlejack, Mexican jackrabbit, Jackass rabbit
HEAD AND BODY LENGTH	20–24 inches (51–61 cm)
TAIL LENGTH	2–3 inches (5–7.5 cm)
BODY WEIGHT	6–13 pounds (2.7–6 kg)

ALASKAN HARE *Lepus othus*	
COMMON NAMES	Arctic hare, Tundra hare, Swift hare
HEAD AND BODY LENGTH	20–23 inches (51–58 cm)
TAIL LENGTH	2–4 inches (5–10 cm)
BODY WEIGHT	8.8–16 pounds (4–7.3 kg)

SNOWSHOE HARE *Lepus americanus*	
COMMON NAMES	Snowshoe rabbit, Varying hare
HEAD AND BODY LENGTH	13–18 inches (33–46 cm)
TAIL LENGTH	1–2 inches (2.5–5 cm)
BODY WEIGHT	2–5 pounds (0.9–2.3 kg)

under brush or rock cover. Drinking is not necessary because they get water from the plants they eat. The pelage on the back is a brownish gray and grades to a lighter buff on the belly. They have large brown eyes that at night reflect a deep ruby red in car headlights. Black-tailed jackrabbits get their common name because the top of the tail is black (unlike white-tailed jackrabbits). The breeding season becomes shorter in the northern portions of the blacktails' range, whereas in the south they breed throughout the year and may produce seven litters annually. Gestation is forty-three days. Litter sizes are over four in northern populations but two or three in southern regions. Given this level of reproductive activity, it is not surprising that population densities can increase dramatically. Like snowshoe hares, black-tailed jackrabbits exhibit cyclic fluctuations in population size, and high densities can decimate agricultural areas. Like other hares, black-tailed jackrabbits are very fast and agile, reaching speeds up to forty miles per hour for short distances. Their erratic, zigzag escape behavior as they dash through brush and over rocks is interspersed with occasional high leaps to get a better view of pursuers. This creates a definite challenge for potential predators such as coyotes, bobcats, eagles, hawks, and owls.

The white-tailed jackrabbit occurs from southern Alberta and Saskatchewan south to New Mexico, and from the Cascade and Sierra Nevada mountains east to Wisconsin. It prefers grassy or brushy habitats at elevations up to thirteen thousand feet. The species is nocturnal although, like other hares, it may feed in the late afternoon or early morning. The whitetail is essentially solitary, although aggregations may occur at preferred feeding sites or temporarily during the breeding season. Similar to the black-tailed jackrabbit in general pelage coloration, it is only slightly larger. However, unlike the black-tailed jackrabbit, the whitetail in northern portions of its range or at higher elevations molts into a white (or pale gray) pelage as do snowshoe and Arctic hares. As expected from the common name, the white-tailed jackrabbit has a white or buff-color tail. It does not compete well with the blacktail, and where both species occur in the same area the whitetail moves to higher elevations. Depending on latitude and elevation, the whitetail produces from one to four litters a year. Four or five young per litter are typical, following a gestation of thirty to forty-three days. Like all hares, the young are precocial and capable of running within an hour after birth. Weaning occurs at one month of age, and the young are independent when two months old.

Antelope jackrabbits range from south central Arizona south through western Mexico. They are larger than black-tailed jackrabbits, the only other hare that shares their range. Pelage color is similar to that of black-tailed jackrabbits except the sides are whitish. The ears are unusually large, seven to eight inches long, and function as radiators to reduce body heat in the desert areas where they live. Unlike black-tailed jackrabbits, the ears have a fringe of white hairs on the sides and do not have black tips. Antelope jackrabbits are probably the fastest hares, with running speeds that approach forty-five miles per hour. They display a white rump patch as they run, similar to that of the pronghorn antelope, which is the basis for their common name. Antelope jackrabbits feed on grasses, desert shrubs such as mesquite, and cactus. They do not drink water but instead meet their water needs with succulent vegetation.

CONSERVATION CONCERNS

Given their extensive range and high reproductive potential, black-tailed jackrabbits and snowshoe hares receive more attention than the other species of hares. Blacktails are not really managed by state or federal resource agencies, however, other than to control population densities. Although they are occasionally hunted for food or sport, blacktails are of most economic importance as serious crop pests and as competitors with livestock for forage. This is especially true when their population densities occasionally reach thousands per square mile. Historically in western states, "rabbit drives" were the solution to these periodic population eruptions. Large groups of ranchers and farmers would surround a tract of land and drive the hares into a corral where several thousand might be killed in an afternoon. Past efforts at predator control throughout the western United States also played a part in increased densities of black-tailed jackrabbits. Today, chemical repellents, fences, and poisons are more commonly used to keep populations in check. High jackrabbit densities can also negatively affect airports, because the hares attract numerous mammalian and avian predators that are then hazardous to aircraft as they land or take off. Snowshoe hares also severely damage commercial timber plantations and orchards when they reach extremely high densities. Application of predator odors is sometimes used in an attempt to reduce damage, as are fences, scare devices, and shooting. Although they do not occur in any national parks, white-sided jackrabbits are considered endangered because of their limited range and population size.

RODENTS

There are more species of rodents throughout the world—about 2,016 recognized species—than in any other mammalian order. Rodents comprise about 43 percent of all living mammalian species. They occur just about everywhere except polar regions and have been introduced into the few places they do not naturally occur. Rodents generally are small animals; most weigh about 0.7 to 1.0 ounce. The world's largest rodent is the South American capybara, which may reach 110 pounds. The largest North American rodent, seen in many national parks, is the beaver. Large adults may weigh seventy-five pounds.

Despite the large number of species and their worldwide geographic distribution, rodents are surprisingly uniform in their physical characteristics. The key characteristic of all rodents is a single pair of upper and lower incisors. These large, sharp, chisellike teeth are used for gnawing—the word *rodent* is derived from the Latin for "to gnaw." Incisors grow throughout an individual's life. All rodents lack canine teeth and have a large gap (called a *diastema*) between their incisors and cheekteeth. Compared to most other mammals, rodents have relatively few teeth, usually no more than sixteen. Most rodents are herbivores, although a few species may eat insects or other animal matter at certain times of the year.

Rodents affect our lives in numerous ways. Worldwide, they consume an average of $30 billion worth of agricultural crops each year. Rodents also are vectors for numerous serious diseases, including bubonic plague, Lyme disease, and hantaviruses. On a positive note, many species of rodents are important in the fur industry, including muskrats and beavers in North America. Rodents also are important in medical research labs and are central to many scientific studies in physiology, ecology, psychology, and other disciplines. Finally, many people keep pet rodents, such as guinea pigs, hamsters, and gerbils.

Because most rodents are small, cryptic, and nocturnal, the vast majority of species will never be seen by park visitors. These fall into the general categories of mice and rats, as well as pocket gophers. A general consideration of these two groups follows. Detailed accounts are provided only for the larger, diurnal species likely to be seen by park visitors—either the animals themselves or their sign—including tree squirrels, ground squirrels, prairie dogs, beaver, muskrat, woodrats, and the porcupine.

Woodrats

The common name "woodrat" clearly has a pejorative connotation because they are thought of as typical brown (Norway) or black rats, which most people abhor. This is unfortunate because woodrats are attractive, harmless, and intriguing animals. Although woodrats are rodents, their large eyes, soft, silky fur, and haired tails all serve to distinguish them from introduced brown or black rats. Ten species of woodrats occur in the United States (nine of them in national parks) with an additional

Desert woodrats are common residents in several western parks but are rarely seen because they are shy, nocturnal rodents.

eleven species farther south in Mexico and Central America. Aside from their geographic ranges and morphology, the various species of woodrats are

BUSHY-TAILED WOODRAT	
Neotoma cinerea	
HEAD AND BODY LENGTH	7–9 inches (18–23 cm)
TAIL LENGTH	5–7 inches (13–18 cm)
BODY WEIGHT	7–20 ounces (198–567 g)

ALLEGHENY WOODRAT	
Neotoma magister	
HEAD AND BODY LENGTH	9 inches (23 cm)
TAIL LENGTH	7 inches (18 cm)
BODY WEIGHT	12.2 ounces (346 g)

DESERT WOODRAT	
Neotoma lepida	
HEAD AND BODY LENGTH	5–8 inches (13–20 cm)
TAIL LENGTH	4–7 inches (10–18 cm)
BODY WEIGHT	4.5–5.6 ounces (128–159 g)

DUSKY-FOOTED WOODRAT	
Neotoma fuscipes	
HEAD AND BODY LENGTH	7–10 inches (18–25 cm)
TAIL LENGTH	6–9 inches (15–23 cm)
BODY WEIGHT	7–12.6 ounces (198–357 g)

EASTERN WOODRAT	
Neotoma floridana	
HEAD AND BODY LENGTH	7–10 inches (18–25 cm)
TAIL LENGTH	5–7 inches (13–18 cm)
BODY WEIGHT	10 ounces (283 g) males
	8.2 ounces (232 g) females

MEXICAN WOODRAT	
Neotoma mexicana	
HEAD AND BODY LENGTH	7–8 inches (18–20 cm)
TAIL LENGTH	4–8 inches (10–20 cm)
BODY WEIGHT	5.2–8.7 ounces (147–247 g)

SOUTHERN PLAINS WOODRAT	
Neotoma micropus	
HEAD AND BODY LENGTH	9 inches (23 cm)
TAIL LENGTH	6 inches (15 cm)
BODY WEIGHT	9 ounces (255 g) males
	8 ounces (227 g) females

STEPHEN'S WOODRAT	
Neotoma stephensi	
HEAD AND BODY LENGTH	7 inches (18 cm)
TAIL LENGTH	5 inches (13 cm)
BODY WEIGHT	4.2–6.3 ounces (119–179 g)

WHITE-THROATED WOODRAT	
Neotoma albigula	
HEAD AND BODY LENGTH	7 inches (18 cm)
TAIL LENGTH	6 inches (15 cm)
BODY WEIGHT	8 ounces (227 g) males
	6.6 ounces (187 g) females

fairly similar in behavior, reproduction, and many other aspects of life history. Bushy-tailed woodrats are the largest and most widely distributed species, both geographically and in the national parks, and are representative of the group.

Dorsal pelage of the bushy-tailed woodrat is a buff gray (the species name *cinerea* is Latin for gray), with a wash of darker guard hairs. The belly and feet are white. It has relatively large hairless ears, and is unique among woodrats in having a bushy, squirrel-like tail. Except for the tail, there are only minor differences in size and pelage color in other species of woodrats. Woodrats are medium-sized rodents and most weigh less than a pound. In some species, males are somewhat larger than females.

Bushy-tailed woodrats occur farther north than any of the other woodrats, from the Yukon south through British Columbia, and most of the western United States east to the Dakotas. Like all woodrats, they are nocturnal, solitary, and active throughout the year. They feed on green parts of herbaceous vegetation, berries, nuts, and occasional insects. Habitats include fissures and ledges among rocks, cliffs, and mountainous woodland areas. Bushy-tails build a cup-shaped or spherical nest from finely shredded vegetation. As with many species of woodrats, the nest is deeply imbedded within a house built from large masses of accumulated sticks tucked into crevices in rimrock, tree cavities, caves, or old buildings.

The other common names for the woodrat are "pack rat" or "trader rat" because it picks up small, shiny, often brightly colored pieces of paper or plastic, tinfoil, bottle caps, feathers, bones, coins, keys, sticks, or other items and incorporates them into the nest. True to its name, it may leave something else in its place as a trade—often the object it was initially carrying. The bushy-tailed woodrat especially exhibits this engaging behavior.

Woodrats do not hibernate and must store large stockpiles of dried vegetation and nuts in their nests to get through a winter. These storehouses (caches) and the area around them are actively defended against other woodrats. Because they are nocturnal and in often-inaccessible habitats, people will rarely see woodrats. Often easily seen, however, are their large stick nests. Also, woodrats in an area urinate and defecate in the same place, called a latrine. After many generations of woodrats use the same latrine site, large encrusted accumulations of crystallized residues of urine and excrement, which may look like mineral deposits, collect and are easily seen on the side of cliffs.

The breeding season of bushy-tailed woodrats

extends through spring and summer, and females may produce two litters a year. An average of three or four pups is born following a gestation of about five weeks. Pups are weaned when a month old and disperse about three months later. Number of litters and average litter size varies geographically but is generally similar among other species of woodrats. Most individuals survive no more than three years. Predators include coyotes, bobcats, weasels, snakes, and owls.

The other species of woodrats in the United States inhabit deserts, woodlands, or rocky upland areas, depending on geographic location and elevation. The Allegheny woodrat was once considered to be a subspecies of the eastern woodrat, but now has species status based on genetic and morphological evidence. It ranges from the Tennessee River northeast to Pennsylvania and New York in rocky bluff habitats. Populations are declining throughout much of their range, although the causes are not certain. The desert woodrat ranges from southern Oregon and Idaho, south through Baja California. This small species is closely associated with rocky sagebrush areas, where it relies heavily on cactus for food, water, and shelter. Southern Oregon south through Baja California is also the range of the dusky-footed woodrat, which uses shrub and woodland habitats. The eastern woodrat occupies a variety of habitats throughout the southeastern United States from Florida west to Colorado and Texas. Unfortunately, populations are declining throughout much of the range. The eastern woodrat builds a stick nest in a wooded area under downed logs, as well as in cracks and crevices of rocky outcrops and cliffs. Similar habitats are occupied by the Mexican woodrat from Colorado and Utah, south through Mexico and Central America. Southern Plains woodrats occur in desert scrub and grasslands from Kansas and Colorado, south through New Mexico and Texas. The geographic range of Stephen's woodrat is restricted to parts of Utah, Arizona, and New Mexico, where it occupies rocky habitats dominated by junipers. White-throated woodrats are widely distributed throughout the southwest United States and Mexico in arid, rocky, and wooded habitats.

CONSERVATION CONCERNS

Population declines in Allegheny and eastern woodrats have been attributed to loss of habitat, diseases, or predation. Monitoring of woodrats in a region by state wildlife resource agencies can help determine population trends, but unless they are believed to be declining, there are no conservation or management initiatives specifically for woodrats. Woodrats often occur in isolated, remnant populations that are prone to dying out (extirpation). Natural recolonization of these sites through dispersal may be impossible because of habitat fragmentation from logging and road construction.

One of the more fascinating aspects of western species, including bushy-tailed, Mexican, desert, and white-throated woodrats, is the use of old nests to determine ecological conditions of regions that were occupied thousands of years ago (a discipline called paleoecology). Huge accumulations of nesting material and associated plants, pollen, bones, invertebrates, and other artifacts collect in very large crevices. These immense "middens" are preserved because of dry conditions and copious amounts of crystallized urine that encrust and stop decay. Examination of accumulated materials can determine ecological conditions up to forty thousand years ago.

Mice and Rats

Numerous species of mice and rats occur in North America and the national parks, including pocket mice, kangaroo rats, kangaroo mice, jumping mice, voles, lemmings, grasshopper mice, deer mice, rice rats, harvest mice, and cotton rats. The familiar brown (Norway) rat and black rat, as well as the house mouse, commonly found in homes, are Old World species that have been introduced widely through-

Although rarely seen, small rodents are the most common mammals throughout the world. Species like this deer mouse are widely distributed throughout North America and occur in most national parks.

out the New World. Although "mice" are generally thought of as smaller than "rats," this is not necessarily the case, and there is no real difference between the terms. Most mice and rats have body and tail lengths of several inches and weigh a few ounces.

Mice and rats are adapted to a wide range of habitats and environmental conditions. They occur from sea level to high elevations and from hot, dry deserts to humid forests, tundra areas, and grasslands. They may eat seeds, grasses, leaves, fruits, and flowers of plants, or be more specialized on other types of vegetation. Species of grasshopper mice eat insects during part of the year. Kangaroo rats, pocket mice, kangaroo mice, and other species live in deserts throughout the western United States. They are highly adapted to harsh conditions of high heat, low precipitation, and limited food resources. Most desert rodents are nocturnal and avoid the heat of the day by remaining inactive in burrows or shelters. They usually do not need free water but get the moisture they need from vegetation. Conserving body water physiologically by reducing evaporative loss and producing very concentrated urine is also important for survival. Whether in deserts, forests, or meadows, park visitors will rarely catch a glimpse of rats or mice. Even if an animal is seen, identification is difficult because many similar-looking species are often in the same area.

Pocket Gophers

Most of the eighteen species of pocket gophers north of Mexico have a very limited geographic range, and only seven species actually occur in national parks (see Table, pages 203–4). Pocket gophers spend almost no time above ground and are rarely seen. What park visitors will see much more often is the evidence of their activity—the extensive mounds and tunnels that result from their burrowing. Gophers have thickset, chunky bodies with short, strong forelimbs and long claws. Most pocket gophers are about twelve inches long and weigh slightly less than one pound. Their eyes and ears are small because they have little need for vision and hearing underground. The front incisors are long and sharp and, with the forelimbs, are used for digging. Incisors protrude such that the mouth can be closed behind them, and a gopher can dig without getting dirt in its mouth. A dominant feature of gophers is fur-lined pouches or "pockets" on each side of the mouth. These are used to transport the roots and tubers on which gophers forage back to underground storage chambers. Burrow systems

can be quite extensive with mounds of excavated dirt throughout. These mounds differ from those of moles in being less conical and more fan-shaped, with coarser dirt. Spreading out from the mounds, one often can see rope-like ridges about two inches wide that mark where dirt has been pushed up by formation of the tunnels.

SQUIRRELS

Squirrels are a family (Sciuridae) of rodents with some of the most familiar and easily seen mammals in the national parks. Sciurids include chipmunks and tree squirrels (well known from suburban backyards), prairie dogs, and ground squirrels. Except for flying squirrels, all are active during the day. Squirrels are very diverse in terms of habitats and socialization. Some are primarily arboreal, including flying squirrels and other tree squirrels. Chipmunks are primarily terrestrial, and prairie dogs, marmots, and ground squirrels are adept burrowers. Prairie dogs are highly colonial, whereas certain tree squirrels are very territorial and intolerant of intruders. The smallest squirrels are the chipmunks and the largest are marmots (also called groundhogs). Regardless of their size, all squirrels are herbivores, with seeds and nuts a primary food source. Certain species, such as the red squirrel, are active throughout the year. Others, especially at higher latitudes and elevations, spend most of the winter in hibernation. During hibernation, metabolism is markedly reduced, such that heart rate, core body temperature, and respiration may be reduced by 90 percent or more. Some species store (cache) food reserves that they use throughout the winter or rely on after emergence from hibernation. We are all familiar with tree squirrels burying nuts in a yard or local park. Other species do not cache food but instead accumulate a large amount of body fat prior to hibernating. Regardless of whether they hibernate, squirrels generally have high mortality rates and form an important part of the prey base for many mammalian and avian predators. Despite heavy predation pressure from terrestrial mammals, hawks, and owls, populations of squirrels are maintained because of fairly high reproductive rates. Several species are popular with hunters, including the familiar gray and fox squirrels. Other species, including the California ground squirrel, can cause significant damage to agricultural areas. Historically, many species of ground squirrels and prairie dogs were eradicated to the extent that they now are threatened or endangered.

CHIPMUNKS

Frisky, vocal, somewhat nervous, these small squirrels are familiar to almost everyone. With a small investment in time and patience in the backyard, people can often get chipmunks to take crackers, nuts, or seeds out of their hands. If not consumed immediately, this food bounty will be stockpiled in the den and the chipmunk will quickly return for another easy meal. The twenty-five species of chipmunks in North America are adapted to life within a variety of habitats, including high elevation meadows, woodlands, rocky slopes, and deserts. Common habitat features include large amounts of ground clutter in the form of vegetation, downed logs, rocks, brush, or stumps. Fences, patios, and hedges provide similar cover in suburban neighborhoods. All chipmunks have short pelage with the familiar alternating dark and light stripes on the back that vary in color and number by species. Pelage color also varies seasonally and in different habitats. The face usually has a single dark stripe through the eye, bordered by white stripes. This characteristic can be used to differentiate between chipmunks and the similar-looking golden-mantled ground squirrel or antelope squirrel. Adding to possible misidentification, most of the sixteen species of chipmunks found within the western national parks look very much alike. Four species are restricted to single parks: the Panamint chipmunk in Death Valley National Park, the red-tailed chipmunk in Glacier National Park, and the Siskiyou and Sonoma chipmunks in Redwood National Park. Many parks, however, such as Sequoia–Kings Canyon and Yosemite, have several species of chipmunks. Throughout much of the eastern United States, only one species occurs—the eastern chipmunk. The remaining twenty-four species of chipmunks occur throughout the west, and in many places telling them apart is practically impossible. Habitats, elevation, and vocalizations may help narrow the choices.

Eastern Chipmunk

This is a common species in deciduous forests, usually where the forest floor has stumps, rocks, and woody debris. Eastern chipmunks also inhabit residential areas, parks, and other suburban areas. The geographic range extends from the East Coast to eastern North Dakota and in the Midwest south to Louisiana, Mississippi, and Alabama.

Burrow entrances usually are associated with rocks, logs, or other ground litter, although eastern chipmunks may use tree cavities as well. Burrows may be short or quite extensive.

Unlike moles and pocket gophers, the chipmunk never leaves evidence of its burrowing in the form of dirt mounds, and entrances and tunnels are always inconspicuous. Chambers in the burrow are used to store seeds and nuts for the winter. The genus name for the chipmunk (*Tamias*) is Greek for "a hoarder" or "one who stores." Stockpiles can be huge—up to a bushel of material. This is necessary for hibernation because, unlike the ground squirrel, the chipmunk does not put on body fat. Food is brought back to the burrow in the cheek pouches. When these pouches are stuffed to capacity, it appears as if a chipmunk has a golf ball on each side of its head. The Eastern chipmunk forages primarily on seeds, as do all chipmunks. Acorns and other hard mast are especially important resources. Chipmunks are opportunistic foragers and eat other foods as they become available seasonally, including berries, mushrooms, flowers, fruits, and invertebrates. Although many chipmunks may inhabit an area and search for food, an individual's core burrow area, including its cache of food, is defended from other chipmunks. Caches represent a great investment in time and energy to accumulate, and must be protected from potential raiders. So chipmunks vigorously defend a territory around their burrows.

Eastern chipmunks are true hibernators, entering torpor in late fall or early winter. Like all mammals, throughout the winter chipmunks periodically rouse from hibernation—and soon reenter—without leaving their dens. They emerge from hibernation in early spring, sometimes even before the snow clears. The breeding season begins soon thereafter. Litter size averages three to four young following a gestation of one month. Neonates are very altricial and remain in the mother's den nursing and gaining weight. Their eyes open when they are a month old. Young emerge from the den for the first time at about six weeks of age and disperse a couple of weeks later. Life expectancy in the wild probably is no more than about a year. Primary predators are hawks, weasels, foxes, raccoons, snakes, and domestic cats.

CONSERVATION CONCERNS
Whether in a national park or their own backyard, most people enjoy watching eastern chipmunks scamper about. These inoffensive squirrels may occasionally damage an ornamental plant by eating the bulb, but in wild habitats are of no significance in terms of conservation or management programs, and are welcome additions to any outdoor experience.

Least Chipmunk

*(Pages 120–21)
Eastern chipmunks
are common and
familiar small
squirrels in forests
and suburban
backyards
throughout the
eastern United
States. Many species
of chipmunks are
recognized by the
stripe along the side
of the body and along
the eye.*

*Of the twenty-four
species of chipmunks
in the western United
States, the least
chipmunk is the most
common and widely
distributed. It occurs
in nineteen national
parks, ranging as far
east as Voyageurs
National Park, where
it overlaps with the
eastern chipmunk.*

The least chipmunk has the widest geographic distribution and occurs in a greater variety of habitats (and more national parks) than any of the other twenty-three species of western chipmunks. The least chipmunk occurs in Canada from the Yukon to Ontario, throughout most of the western United States, and northern Minnesota, Wisconsin, and Michigan. Its range overlaps those of ten other western species of chipmunks. The least chipmunk occupies dry sagebrush desert habitats, rocky areas, montane forests, and high elevation alpine tundra. As might be expected from the common name, the least chipmunk is the smallest of all chipmunks. In most ways, however, its general life history charac-

teristics and those of the other western species are quite similar to those of the eastern chipmunk.

Because habitats often lack cover, the least chipmunk is fairly secretive to avoid the array of hawks, snakes, and terrestrial predators that can take it as it forages during the day. Unlike the eastern chipmunk, its tail is held straight up as it runs. As with other chipmunks, seeds and mast are preferred, but it will eat a variety of other vegetation and insects. The cheek pouches of the least chipmunk are often bulging with seeds, which are transported back to the storage den in an excavated tunnel or hollow log. In hot, arid environments it may go into torpor during the summer and reduce activity above ground. Hibernation begins in late fall or early winter. The winter den is lined with shredded bark, fur, or feathers, and has a large store of food. Breeding is in March or April, soon after emergence from winter hibernation. Gestation is a month and litter size averages four to six young. Each tiny altricial newborn is two inches long and weighs about 0.7 ounce. Young nurse for a month or more, grow rapidly, and like eastern chipmunks, disperse by two months of age.

CONSERVATION CONCERNS

The least chipmunk, like the eastern chipmunk, is abundant throughout its range and is of no concern in terms of conservation issues or management programs but is appreciated for the aesthetics it provides. Least chipmunks, and all the other species, provide a service in the number of their cached seeds that remain uneaten and eventually sprout. This aids natural reforestation. In many national parks, least chipmunks lose their natural shyness, and large numbers (referred to by one authority as "peanut populations") may gather in response to people feeding them at campsites. Two other western species—Palmer's chipmunk (restricted to the Spring Mountains of southwestern Nevada) and the gray-footed chipmunk (found in Carlsbad and Guadalupe Mountains National Parks)—are uncommon and currently listed as threatened.

EASTERN CHIPMUNK *Tamias striatus*	
HEAD AND BODY LENGTH	6–7 inches (15–18 cm)
TAIL LENGTH	3–4 inches (8–10 cm)
BODY WEIGHT	3–4 ounces (85–113 g)

LEAST CHIPMUNK *Tamias minimus*	
HEAD AND BODY LENGTH	4 inches (10 cm)
TAIL LENGTH	4 inches (10 cm)
BODY WEIGHT	1–2 ounces (28–57 g)

MARMOTS

Commonly called groundhogs, marmots are the largest of all squirrels. There are six North American species of marmots. All but Vancouver marmots occur in at least one national park and are easily seen when active. All marmots are chunky and heavy-bodied, with short necks, small rounded ears, and short, powerful forelimbs built for burrowing. Pelage color varies among species—and geographically within species—from brown to reddish orange to gray. Fur is generally long, coarse, and appears somewhat grizzled.

The only species of marmot in the eastern United States is commonly known as the wood-chuck. It has a widespread distribution in eastern Alaska, in southern Canada from British Columbia to Quebec, and from Minnesota and Mississippi eastward except for Florida. Woodchucks have become very common because of forest clearing and associated agriculture. They inhabit open wood-lots, meadows, and pastures, with cover provided by fences, boulders, or downed logs. They are commonly seen waddling slowly along as they forage on grasses, forbs, or cultivated plants. Burrows are built in well-drained soils. They are extensive, often have a number of chambers, and are marked by a large pile of dirt at the entrance. Woodchucks can often be seen sitting on the dirt pile to gain a better look at their surroundings as they watch for coyotes, foxes, eagles, large hawks, and other predators.

The yellow-bellied marmot is a common resident throughout the Rocky Mountain region. Often called a rockchuck, this large squirrel inhabits remote rocky habitats in several national parks.

Foraging throughout summer and autumn is essential to gaining weight prior to winter. The life history of woodchucks and other marmots revolves around burrowing and winter hibernation. Males are larger than females, but body weight of marmots is highly variable seasonally because they must put on a great deal of fat. Unlike chipmunks, woodchucks don't store food in their burrows, and body weight may double prior to entering hibernation in late autumn. Marmots are true hibernators and may spend up to six months underground in torpor. Their body temperature may decrease to a few degrees above ambient, and heart rate and oxygen consumption decrease to 10 percent or less of active levels. Male woodchucks emerge from hibernation first, and breeding begins when females arouse in March or April. Gestation is about thirty days, and litters average four to five altricial young. Neonates are only about four inches long and weigh 0.9 ounce, but they grow rapidly. Weaned by six weeks of age, the young disperse about two weeks later.

Like woodchucks, yellow-bellied marmots are geographically widespread. The most southerly of the western marmots, they occur in the Cascade and Rocky Mountain regions from southern British Columbia and Alberta south to California and New Mexico. They den in rocky areas (their other common name is rockchuck) with suitable vegetation for foraging. There is an isolated population in the Black Hills of South Dakota. Hoary marmots extend throughout Alaska, western Canada, Montana, Idaho, and Washington. Colonies occupy high elevation meadows with abundant forage and boulders to conceal burrowing. The Alaska marmot is restricted to the Brooks Range of northern Alaska, where it inhabits rocky slopes near prime foraging areas, which sit atop permanently frozen soil. Prime predators include grizzly bears, wolverines, and wolves. Two other species have very restricted ranges. The Vancouver marmot is found only on Vancouver Island, British Columbia. Likewise, the Olympic marmot is only on Washington's Olympic Peninsula, where individuals form large colonies.

CONSERVATION CONCERNS

Woodchucks gain the national spotlight every February 2 on Groundhog Day as "Punxsutawney Phil" emerges from his den to prognosticate the weather for the next six weeks. For the remainder of the year, however, woodchucks are much less beloved. They damage agricultural fields by their burrowing and direct consumption of crops, and sometimes damage gardens by foraging or trampling. Their large burrow entrances create a hazard for horseback riders and livestock. On the positive side, they turn over and aerate a tremendous amount of soil throughout their range, and abandoned dens provide shelter for numerous other species. Marmots are not game animals, but many hunters shoot them for sport using small caliber rifles. Shooting and placing poison in burrows are the most effective ways to reduce populations in agricultural areas. However, these are short-term remedies because other animals quickly appear to fill the void. Hoary and yellow-bellied marmots are generally found in wilderness areas away from human development and are less often considered pests. Populations of woodchucks and other marmots in the United States are stable and usually of little concern to conservationists or resource managers. Vancouver marmots, however, are endangered.

WOODCHUCK *Marmota monax*	
COMMON NAMES	Ground hog, Whistle pig
HEAD AND BODY LENGTH	16–20 inches (41–51 cm)
TAIL LENGTH	5–6 inches (13–15 cm)
BODY WEIGHT BEFORE HIBERNATION	10 pounds (4.5 kg)
BODY WEIGHT AFTER HIBERNATION	7.5 pounds (3.4 kg)

YELLOW-BELLIED MARMOT *Marmota flaviventris*	
COMMON NAMES	Rockchuck
HEAD AND BODY LENGTH	13–19 inches (33–48 cm)
TAIL LENGTH	5–8 inches (13–20 cm)
BODY WEIGHT BEFORE HIBERNATION	7.5 pounds (3.4 kg)
BODY WEIGHT AFTER HIBERNATION	5.3 pounds (2.4 kg)

ALASKA MARMOT *Marmota broweri*	
COMMON NAMES	Brooks Range marmot
HEAD AND BODY LENGTH	18–22 inches (46–56 cm)
TAIL LENGTH	7–10 inches (18–25 cm)
BODY WEIGHT BEFORE HIBERNATION	12 pounds (5.4 kg)
BODY WEIGHT AFTER HIBERNATION	7 pounds (3.2 kg)

HOARY MARMOT *Marmota caligata*	
HEAD AND BODY LENGTH	18–22 inches (46–56 cm)
TAIL LENGTH	7–10 inches (18–25 cm)
BODY WEIGHT BEFORE HIBERNATION	14 pounds (6.3 kg)
BODY WEIGHT AFTER HIBERNATION	7 pounds (3.2 kg)

OLYMPIC MARMOT *Marmota olympus*	
HEAD AND BODY LENGTH	19–21 inches (48–53 cm)
TAIL LENGTH	7–10 inches (18–25 cm)
BODY WEIGHT BEFORE HIBERNATION	15.6 pounds (7.1 kg)
BODY WEIGHT AFTER HIBERNATION	7.5 pounds (3.4 kg)

GROUND SQUIRRELS

Ground squirrels are one of the more ubiquitous and easily seen groups of mammals throughout national parks. These diurnal herbivores, members of the squirrel family (Sciuridae), are common around campgrounds in many parks and sometimes approach visitors for handouts. There currently are twenty-one species in the United States called ground squirrels, several of which have been recognized only in the last twenty years. Of these, sixteen species occur in at least one of the national parks. In addition, there are five species of antelope ground squirrels in North America, three of which are in national parks. All ground squirrels and antelope squirrels exhibit fairly similar morphological and life history characteristics. Three species representative of ground squirrels are discussed here because they are widely distributed geographically and are the most likely to be seen in the national parks.

White-tailed Antelope Squirrel

This fascinating squirrel is found in arid, rocky areas with generally sparse, shrubby vegetation throughout the southwestern United States from southeast Oregon to Arizona and New Mexico, south through Mexico and Baja California. Dorsal pelage is a brownish gray with white throat and belly. A distinct white stripe on each side extends from the shoulders over the hips. The underside of the bushy tail is white and easily seen because, unlike other ground squirrels, the antelope squirrel characteristically holds its tail vertically over its back. Smaller than ground squirrels, the antelope squirrel is larger than the chipmunk and does not have facial stripes.

The white-tailed antelope squirrel is unusual among small desert rodents in that it is active during the day and throughout the year, not hibernating like other ground squirrels. Most small animals avoid the blistering daytime heat of the desert by remaining in burrows and foraging only at night. But the antelope squirrel is extremely well adapted to hot, arid conditions and can withstand body temperatures well above 100°F without raising its metabolic rate. It will drink if water is available but can survive on the moisture in the vegetation it eats. It also produces highly concentrated urine

that helps save water, and holding its tail over its back helps shade it as the white underside reflects solar radiation. Finally, if temperatures get too high, a white-tailed antelope squirrel will use its forepaws to spread saliva over its head to cool itself through evaporation, and ultimately will retreat to a burrow.

Succulent vegetation and seeds or fruits of desert plants, as well as insects, make up the diet. Excess food is carried in cheek pouches and cached in the burrow, which has no telltale mounds at the entrances. White-tailed antelope squirrels are rather poor excavators and for their dens use rocky areas or burrows of other animals such as kangaroo rats.

Mating begins in February. Following a one-month gestation, an average litter of eight young is born in the spring. Larger litters occur if green vegetation is plentiful. Young may appear aboveground a week or more before they are weaned at two months of age.

CONSERVATION CONCERNS

Antelope ground squirrels are not colonial and do not form large population aggregations. Although they are captivating species that readily may be seen in national parks as they scurry along with their tails twitching over their backs, antelope squirrels are of no economic significance. Although populations of the white-tailed antelope ground squirrel are secure, Nelson's antelope squirrel, which is restricted to the San Joaquin Valley of California, is a state threatened species and is considered endangered by the World Conservation Union.

Uinta ground squirrels are in several national parks. They are diurnal, active above ground, and easily seen. Like numerous other species of ground squirrels throughout North America, they are a critical part of the ecosystem and a valuable prey base for predators.

Golden-mantled Ground Squirrel

Most likely to be misidentified as a chipmunk, this commonly observed handsome ground squirrel has a broad white stripe on each side of its back, bordered by narrower black lines. It has no facial stripes. Other than the thirteen-lined ground squirrel, and the closely related Cascade golden-mantled ground squirrel, the golden-mantled ground squirrel is the only member of the genus with stripes. Attractive and bright golden-brown pelage on the head and shoulders forms the "mantle" of the common name. The back and tail are a yellow brown; the belly is a dull white, and the dark eyes have a white ring. The Cascade golden-mantled ground squirrel looks the same, but is now considered to be a separate species that does not overlap in distribution. The golden-mantled ground squirrel inhabits spruce, pine, and fir forests in mountainous and valley regions throughout western North America from British Columbia and Alberta south to New Mexico and Arizona.

It is active throughout the day, foraging on a variety of seeds. (The genus name of all ground squirrels, *Spermophilus*, means "seed lover.") Nuts, fruits, fungi, green vegetation and flowers, bird eggs, and insects round out the diet. It carries food back to the den in internal cheek pouches and stores it. Burrow systems are extensive, although not necessarily deep underground, with entrances hidden under fallen logs, rocks, stumps, or brush. As winter approaches, the golden-mantled ground squirrel puts on extensive amounts of body fat in preparation for hibernation, which can extend from late October until April or May. Unlike many other ground squirrels, it is not colonial or social. Numerous mammalian carnivores, hawks, and snakes prey on golden-mantled ground squirrels.

Breeding takes place following emergence of females from hibernation in late spring. Gestation extends for about twenty-seven days, and the average litter size is five. The young are weaned at about a month of age when they begin to appear above ground. Young leave the natal area by late July before they are full grown.

CONSERVATION CONCERNS

Unlike other species of ground squirrels that cause problems in agricultural areas, golden-mantled ground squirrels are not found in developed areas. As such, they are of no economic significance and of little concern to conservation or management programs. Golden-mantled ground squirrels are commonly encountered around campsites throughout the national parks, where they can be shameless beggars and are a favorite of visitors because of their endearing behavior and conspicuous coloration. Unfortunately, the squirrels can carry the flea-borne plague bacterium and may be a focus of the disease.

Rock Squirrel

As expected from the common name, this squirrel inhabits rough and rocky terrain of canyons and hills throughout the west from northern Colorado and Utah to Texas and south through Mexico. It is large for a ground squirrel, with a bushy tail similar to that of a tree squirrel. The dorsal pelage is a mottled, speckled, multicolored black, brown, and gray (which gives rise to the species name) with variable amounts of black on the head and shoulders. The belly coloration also varies from a buff white to reddish brown. There is a pronounced white eye ring. No other ground squirrel within its range is similar to the rock squirrel.

Rock squirrels burrow with den entrances under rocks, roots, or debris that offer good lookout posts of the surrounding area. Like other ground squirrels, they forage during the day for seeds, fruits, green vegetation, eggs, insects, and carrion. They also may climb trees for acorns and juniper berries, which

is unusual for ground squirrels. Unlike the golden-mantled ground squirrel, rock squirrels are colonial and somewhat more social. A colony consists of a dominant male, several subordinate males, and females. Predators include the usual array of mammalian carnivores, hawks, and snakes, but predation does not have a major effect on populations.

Like all ground squirrels, rock squirrels are obligate hibernators with associated annual fat cycles. Prior to entering hibernation, an individual may weigh twice what it does upon emergence. Females breed after emerging from hibernation in the spring. Following a gestation of four weeks, litters are dropped between May and early July. Average litter size is four or five altricial young. Pups emerge from the burrow when about two months old and disperse about a month later.

CONSERVATION CONCERNS

Certain species, especially the California ground squirrel and Belding ground squirrel, have long been considered by farmers to be serious crop depredators. In fact, crop damage by ground squirrels may approach $30 million annually. Rock squirrels eat fruit and can harm orchards because of their climbing ability. Historically, management of ground squirrels involved control measures such as poison baits, trapping, shooting, anticoagulants, and fumigants to reduce populations. Eradication efforts were so successful that many species and subspecies are now considered endangered, threatened, or of conservation concern. As such, if control is necessary, it should be applied diligently on a site-specific basis. Populations of rock squirrels are considered stable throughout their range, however.

WHITE-TAILED ANTELOPE SQUIRREL *Ammospermophilus leucurus*	
MEAN HEAD AND BODY LENGTH	6 inches (15 cm)
TAIL LENGTH	2.4 inches (6 cm)
BODY WEIGHT	4 ounces (113 g)

GOLDEN-MANTLED GROUND SQUIRREL *Spermophilus lateralis*	
MEAN HEAD AND BODY LENGTH	7 inches (18 cm)
MEAN TAIL LENGTH	3.5 inches (9 cm)
BODY WEIGHT	6 ounces (170 g) spring
	12 ounces (340 g) autumn

ROCK SQUIRREL *Spermophilus variegatus*	
MEAN HEAD AND BODY LENGTH	11 inches (28 cm)
MEAN TAIL LENGTH	8 inches (20 cm)
BODY WEIGHT	1 pound (454 g) spring
	2 pounds (908 g) autumn

PRAIRIE DOGS

Although they certainly inhabit prairies, the five species of North American rodents called prairie dogs obviously are not related to dogs. The common name dates to the expedition of Lewis and Clark, who first described a "barking squirrel"—and even sent a live specimen back to Thomas Jefferson. Four of these species—all except the Mexican prairie dog—occur north of Mexico and inhabit the national parks.

Black-tailed prairie dogs have the largest and most easterly distribution of these species, and once ranged from extreme southern Canada, Montana, and the Dakotas, south through western Texas, New Mexico, and Arizona. White-tailed prairie dogs occur from southern Montana to northwestern Colorado. Gunnison's prairie dogs are in Utah, Colorado, New Mexico, and Arizona. Utah prairie dogs are found only in southwestern Utah, northwest of Gunnison's and southwest of white-tailed prairie dogs. There is very little overlap of geographic ranges in the four species, and today all inhabit remnants of their historical distributions.

Prairie dogs are chunky, ground-dwelling rodents closely related to ground squirrels. Different species of prairie dogs look much alike and share many of the same life history characteristics. They have large, dark eyes, short, round ears, and long claws for burrowing. The coarse pelage is generally buff brown to gray, although it may be more reddish brown in Utah prairie dogs. Because they spend so much time underground, pelage often takes on the color of the soil in which they occur. Tails are short and somewhat flat and, as their name suggests, the black-tailed prairie dog has a black-tipped tail. The white-tailed and the smaller Utah prairie dog have shorter tails without the black tip, and these two have a black line over the eyes. Gunnison's prairie dog also has a tail without the black tip but lacks the dark line above the eyes.

Black-tailed prairie dogs inhabit semiarid, short-grass prairies at lower elevations. They avoid tall grass and actually clip vegetation so that it does not obstruct their view of the surroundings. Other species of prairie dogs are most often in high meadows at elevations up to twelve thousand feet. All prairie dogs are diurnal and herbivorous, the diet changing seasonally from grasses, forbs, and herbs in the spring, to seeds in the summer, and roots and remaining green vegetation later in the year. Invertebrates such as grasshoppers or beetles may augment the diet. During stressful winter weather when forage is scarce, prairie dogs may reduce aboveground activity and remain for extended periods of time in their burrows. White-tailed, Utah, and Gunnison's prairie dogs all hibernate, but black-tails accumulate fat reserves and remain active throughout the year. Prairie dogs are best known for being highly social and forming large colonies, or "towns." Black-tailed prairie dogs are the most social. Historically, colonies tens of thousands of square miles in size containing hundreds of millions of black-tailed prairie dogs extended across the plains. Other species form much smaller colonies. Within larger colonies, there are smaller social subunits called "wards" or "precincts," and within these are extended family groups called "coteries." A coterie consists of an adult male, several adult females, and their yearlings and juveniles. The coterie protects the territory immediately surrounding its burrow system from unrelated prairie dogs.

As part of the complex social system, familiar behaviors include "kissing," sniffing, playing, mutual grooming, and various postures. Vocal communication also is well developed in all prairie dogs

Prairie dogs are colonial ground squirrels with complex social behaviors. Unfortunately, the geographic distributions of all five species, including this white-tailed prairie dog, have been dramatically reduced because of ill-advised population control programs.

BLACK-TAILED PRAIRIE DOG *Cynomys ludovicianus*	
HEAD AND BODY LENGTH	11.4 inches (29 cm)
TAIL LENGTH	3.5 inches (9 cm)
BODY WEIGHT	31 ounces (879 g)

GUNNISON'S PRAIRIE DOG *Cynomys gunnisoni*	
HEAD AND BODY LENGTH	11.4 inches (29 cm)
TAIL LENGTH	2.4 inches (6 cm)
BODY WEIGHT	26 ounces (737 g)

UTAH PRAIRIE DOG *Cynomys parvidens*	
HEAD AND BODY LENGTH	10 inches (25 cm)
TAIL LENGTH	2 inches (5 cm)
BODY WEIGHT	20 ounces (567 g)

WHITE-TAILED PRAIRIE DOG *Cynomys leucurus*	
HEAD AND BODY LENGTH	11.8 inches (30 cm)
TAIL LENGTH	2 inches (5 cm)
BODY WEIGHT	37 ounces (1049 g)

and includes bark-like alarm calls (the genus name means "dog mouse"), chirping, and threat vocalizations. Burrow systems are large, deep, extensive, and well developed. They serve not only as protection against inclement weather, but also help shield prairie dogs from numerous terrestrial and avian predators. Prairie dogs are often seen perched on the top of two-foot-high mounds of excavated dirt, on guard and ready to give the alarm call if a predator is sighted. Prior to their near extinction, black-footed ferrets were prime predators on prairie dogs. Today, badgers, coyotes, snakes, large hawks, and golden eagles are the major threats.

Male prairie dogs and female blacktails become sexually mature at about two years of age. Females of other species may breed as yearlings. Breeding occurs in late winter. Gestation is about twenty-nine days in Utah and Gunnison's prairie dogs, and about thirty-five days in the other species. Average litter size is three or four altricial young. Newborns weigh about 0.6 ounce and nurse for five to seven weeks. Interestingly, female black-tailed, Gunnison's, and Utah prairie dogs nurse communally, with mothers nursing juveniles that are not their own. Also, these species exhibit infanticide—killing and eating juveniles. They also cannibalize other prairie dogs that die. Young remain with their parents until they are breeding age.

CONSERVATION CONCERNS
The geographic distributions of all species of prairie dogs are dramatically reduced from historical levels. For example, black-tailed prairie dogs now inhabit less than 2 percent of their former range. Ranchers considered prairie dogs competitors with livestock for forage and the cause of degraded ranges. Federal, state, and local efforts to eradicate prairie dogs involved shooting and poisoning millions of animals. Prairie dog populations also suffered as habitat was lost to agriculture. Population densities and distributions were further reduced because of periodic natural outbreaks of bubonic plague. Colonies today are small, isolated, and scattered throughout the historical ranges. These remnants are often at risk of extirpation from various local threats. All species in the United States are now rare. The Utah prairie dog is a federally threatened species, and other species may eventually be listed. Mexican prairie dogs are endangered. Today, we recognize that prairie dogs do not cause poor range conditions. Their presence is more likely to be symptomatic of overuse. Such realization may have come too late. Maintaining vestiges of prairie dogs and their ecosystems presents a major conservation challenge for the future.

TREE SQUIRRELS

As the name implies, tree squirrels are closely associated with forest habitats, and include the familiar gray and fox squirrels, red squirrels, and the common but seldom-seen flying squirrels. Other, more geographically restricted species of tree squirrels also occur on national parks. Abert's squirrels (also called Kaibab squirrels) are in Grand Canyon, Mesa Verde, and Saguaro National Parks, Arizona gray squirrels are in Saguaro, and introduced red-bellied squirrels are in Biscayne National Park.

Fox and Gray Squirrels

Eastern gray squirrels and fox squirrels are probably the most commonly observed species of mammals in the eastern United States. Both occur from the Atlantic Coast through the midwestern states. Distribution of eastern gray squirrels extends further into the northeastern states; fox squirrels are found further west to Montana, Wyoming, and Colorado. A third species, the western gray squirrel, is distributed along coastal and inland mixed woodlands from northern Washington to southern California. Gray squirrels, as their name suggests, have silver-gray or dark gray-brown dorsal pelage with a white chin, belly, and eye ring. The edges of the long bushy tail are white or silver. Fox squirrels are highly variable in pelage color, but often are a grizzled brownish red above, with a distinct orange wash on the belly and under the tail. Their tails are edged in a rusty orange color, and they have a tan eye ring. All-black (melanistic) individuals occur in gray and fox squirrels, as do occasional albinos. While sitting, gray and fox squirrels often hold their tails in a characteristic S-shape above the back (the genus name *Sciurus* is derived from two Greek words meaning "shade tail"). Besides pelage color, eastern gray squirrels differ from fox squirrels by their smaller size.

Fox and gray squirrels are closely associated with mature hardwood or mixed hardwood and conifer forests, where they feed primarily on cones, acorns, and other mast including beechnuts, walnuts, and hickory nuts. Mushrooms, berries, grain, fruits, insects, and flowers round out the diet, especially in spring and summer when mast is not yet available. All three species can be seen during summer and fall as they bury nuts, apparently randomly, about an inch deep in the ground. This "scatter hoarding" ensures that a competitor cannot steal all the food stored in one place. Squirrels use their memory and sense of smell to locate buried nuts, although they fail to retrieve some of them. Large trees not only provide food and travel corridors but furnish shelter as well. Tree squirrels build two kinds of nests. Large, globular, well-insulated aggregations of sticks and leaves are readily seen on branches seventy feet high or more. These nests are especially noticeable after leaves fall. Dens also are constructed in natural tree cavities. Cavities must be large enough to accommodate the squirrels and oriented so they remain dry during rainfall. Tree squirrels are active during the day. They do not hibernate during the winter, although they remain in dens during bad weather. Gray squirrels seem to prefer woodlands with denser understory, and fox squirrels spend more time on the ground and prefer more open areas between large trees. Fox and eastern gray squirrels also occur in residential areas where there are large trees and sufficient habitat to support them. Western gray squirrels are much less tolerant of people and developed areas, however, and are generally more cautious. Tree squirrels are preyed upon by a variety of hawks, owls, and terrestrial carnivores including coyotes, raccoons, bobcats, martens, and domestic dogs and cats. Compared to food shortages, disease, and inclement weather, however, predators have a minor impact on populations.

There are two peaks of reproductive activity in eastern gray and fox squirrels. Breeding occurs in January and February, with a secondary peak in June and July. Depending on age and body condition, adult females produce either one or two litters per year. Gestation is forty-four days. Average litter size is two or three altricial young, and pups weigh about 0.5 ounce at birth. Weaned by eight to ten weeks of age, the pups are independent soon afterwards. Dispersing young and adults during the autumn often results in movement of a large number of squirrels, a phenomenon called the "fall reshuffle." Western gray squirrels have breeding peaks in spring and summer and the same gestation period, but adult females generally have only one litter per year.

CONSERVATION CONCERNS

Gray and fox squirrels are both very popular game animals, with an estimated forty million harvested annually during the fall hunting season. These species also contribute to tree regeneration because of the acorns and other nuts that they bury and never retrieve that sprout seedlings. Besides hunting, tree squirrels also are a very important nonconsumptive resource for people to view and photograph. Because people like to see squirrels in parks, eastern gray and fox squirrels have been introduced into several areas in the western United States. Unfortunately, however, these introduced species appear to displace the native western gray squirrels where they overlap because they are more aggressive and have a higher reproductive potential.

EASTERN GRAY SQUIRREL *Sciurus carolinensis*	
COMMON NAMES	Cat squirrel, Black squirrel, Silvertail
HEAD AND BODY LENGTH	9–11 inches (23–28 cm)
TAIL LENGTH	6–9 inches (15–23 cm)
BODY WEIGHT	0.9–1.5 pounds (0.4–0.7 kg)

FOX SQUIRREL *Sciurus niger*	
COMMON NAMES	Black squirrel, Red squirrel, Stump-eared squirrel
HEAD AND BODY LENGTH	10–15 inches (25–38 cm)
TAIL LENGTH	9–14 inches (23–35 cm)
BODY WEIGHT	1.5–2.6 pounds (0.7–1.2 kg)

WESTERN GRAY SQUIRREL *Sciurus griseus*	
COMMON NAMES	Columbian gray squirrel, Silver gray squirrel
HEAD AND BODY LENGTH	10–15 inches (25–38 cm)
TAIL LENGTH	9–14 inches (23–35 cm)
BODY WEIGHT	1.1–2.2 pounds (0.5–1.0 kg)

Red and Douglas's Squirrels

Highly vocal and confrontational, both these species of small tree squirrels are likely to make themselves heard if you are in "their" section of the woods. Red squirrels are the most widely distributed member of the squirrel family (Sciuridae) in North America. They range from Alaska through Canada and the Rocky Mountains south to Arizona and New Mexico, in north-central states, and in the east from Maine south along the Appalachian Mountains to Georgia. Closely related Douglas's squirrels have a much more limited distribution from southern British Columbia south through California. Life history patterns are similar in both species (some authorities have suggested that they actually are the same species), and they also mirror other tree squirrels in many ways.

Red squirrels are somewhat misnamed. The dorsal pelage is more of a rich chestnut brown. The belly fur is white as is the eye ring, and during the summer

there is a distinct lateral black line between the upper and lower pelage. Douglas's squirrels are similar except the belly fur and eye ring are buff-orange color.

Preferred habitats are coniferous forests, primarily pine, spruce, hemlock, and fir, although they also occur in mixed and deciduous woodlands. Like gray and fox squirrels, red and Douglas's squirrels nest in cavities, often enlarged woodpecker holes. Leaf and stick nests about twelve inches in diameter (smaller than those of gray or fox squirrels) also are constructed on tree limbs near the trunk. Conifer seeds and nuts of deciduous trees are favored foods, with a variety of mushrooms (some poisonous to people), insects, buds, and small vertebrates rounding out the diet. Red and Douglas's squirrels significantly differ from the much larger gray and fox squirrels in that they do not bury seeds and nuts randomly but instead "larder hoard" their cache in one or more piles (called middens). These stockpiles are aggressively defended from intruders, and are the reason red and Douglas's squirrels are so vocal. Intruders to the territory will elicit a loud, aggressive chat-

tering, tail-flicking, feet-stomping display from the highly agitated resident as it defends its accumulated hoard. Middens may be used by several generations of squirrels, and accumulations of cones and associated debris often become extremely large, encompassing many bushels of material. Both species are diurnal and active throughout the year. Because

DOUGLAS'S SQUIRREL	
Tamiasciurus douglasii	
COMMON NAMES	Chickaree, Spruce squirrel
HEAD AND BODY LENGTH	7–8 inches (18–20 cm)
TAIL LENGTH	4–6 inches (10–15 cm)
BODY WEIGHT	3–11 ounces (85–312 g)

RED SQUIRREL	
Tamiasciurus hudsonicus	
COMMON NAMES	Pine squirrel, Barking squirrel, Boomer, Chickaree
HEAD AND BODY LENGTH	7–8 inches (18–20 cm)
TAIL LENGTH	4–6 inches (10–15 cm)
BODY WEIGHT	3–11 ounces (85–312 g)

The red squirrel, and the closely related Douglas's squirrel, occurs on national parks throughout the western and northern United States. Active throughout the year, this red squirrel is highly vocal and aggressive defending its food cache.

A tree cavity provides warmth and protection for young red squirrels until the litter can leave the nest. These pups will continue to nurse until they are about ten weeks of age.

they do not hibernate, they need the cached seeds in middens to make it through the winter.

The breeding season may begin in late winter and extend until early autumn, although most females have a single litter. Courtship involves frenzied chasing and vocalizing, as individual females are receptive for only one day. Following a thirty-five-day gestation, three to four altricial young are born. Newborns weigh about 0.3 ounce and, as with all squirrels, development is slow. Pups first leave the nest at seven weeks of age and are weaned three weeks later.

CONSERVATION CONCERNS

Although they may strip bark from conifer plantations, damage orchard crops, and occasionally inhibit natural regeneration of some conifers, these beautiful little squirrels are of relatively minor economic or conservation concern. Red squirrels are trapped as a furbearer in some areas. As a game species, however, neither red nor Douglas's squirrels are as important to wildlife management and conservation programs as are gray and fox squirrels.

Flying Squirrels

Like red squirrels, northern flying squirrels occur in coniferous and mixed deciduous forests in Alaska, throughout Canada, and in northern tier states in the east and west with southern extensions in the Rocky Mountains and Appalachians. Deciduous oak/hickory forests throughout the eastern United States are home to the smaller southern flying squirrel. Flying squirrels occur in several national parks, but because they are our only nocturnal squirrels, few park visitors will ever see these beautiful little animals. Flying squirrels have large black eyes and very dense, silky fur—brown on the upper surface with a white belly. The common name is a misnomer in that these species do not fly (bats are the only mammals capable of true flight). Northern and southern flying squirrels are well adapted for gliding, however. They have an extensive furred flap of skin (called a *patagium,* or gliding membrane) between their front and hind limbs. This membrane increases the surface area and allows individuals to glide great distances and be highly maneuverable. Also, their flattened tail is used as a rudder while in the air. Gliding distance may reach 270 feet, although an average glide probably is about sixty

NORTHERN FLYING SQUIRREL *Glaucomys sabrinus*	
HEAD AND BODY LENGTH	6–7.5 inches (15–19 cm)
TAIL LENGTH	5–6 inches (13–15 cm)
BODY WEIGHT	2.5–5 ounces (71–142 g)

SOUTHERN FLYING SQUIRREL *Glaucomys volans*	
HEAD AND BODY LENGTH	4.7–5.5 inches (12–14 cm)
TAIL LENGTH	3–5 inches (8–13 cm)
BODY WEIGHT	1.6–3 ounces (45–85 g)

feet. Flying squirrels glide from high on one tree to a lower height on a second tree, then climb back to the top.

Flying squirrels generally forage for nuts in trees, but also take a variety of plants and fungi on the ground. They are especially fond of tree sap. Unlike other tree squirrels, flying squirrels do not store food for the winter, nor do they hibernate. Activity can be greatly reduced during severe winter weather, however, and they may nest communally to conserve body heat. They nest in tree cavities, vines, or dense matted vegetation. Because they are nocturnal, flying squirrels are a major prey base for a number of different species of owls. Weasels, raccoons, and opossums also take them in their nests. Flying squirrels breed in the spring or early summer and older females may produce two litters per year. Gestation is forty days and average litter size is three or four. Newborns are altricial and weigh only about 0.1 ounce. They grow quickly and increase their body weight by fifteen times when weaned at six weeks of age. By three months of age, they begin their first glides.

CONSERVATION CONCERNS

Aside from being an intriguing part of the mammalian fauna, flying squirrels are of no commercial importance as a furbearer (their fur is much too delicate), nor do they damage tree seedlings, agricultural crops, or residential landscaping.

MOUNTAIN BEAVER

Based on skull characteristics, anatomy, and physiology, mountain beavers are the most primitive living rodents, with a fossil history dating back sixty million years. They have no close living relatives and are the sole living member of the family Aplodontidae. Mountain beavers also are called "boomers," "whistlers," and "sewellels." Mountain beavers do not generally inhabit mountainous regions, nor are they really beavers. They occur only in North America, where they are restricted to the Pacific Northwest from southern British Columbia to California, from the coast east to the Cascade Mountains, and in the Sierra Nevada Mountains of California. They inhabit humid, low elevation areas with dense understory, usually near a stream or in areas with high annual precipitation. Mountain beavers are restricted to wet, humid areas because their kidneys are inefficient and cannot produce concentrated urine. They lose copious amounts of water through urination. Daily production of urine equals 25 to 33 percent of body weight, and individuals must consume large amounts of free water or succulent vegetation.

Mountain beavers look like chunky, overstuffed gophers. The pelage is dark reddish brown to black with a distinct white spot below each ear. They are herbivores and consume numerous different plant species, many of which are unpalatable or even poisonous to other wildlife, such as bracken fern, rhododendron, devil's club, and stinging nettle. They cut much more vegetation than they eat; excess material results in "haymaking"—fresh vegetation stacked two feet high near entrances to burrows. Some of this vegetation is used for nesting material and some is eaten during the winter.

Mountain beavers burrow extensively, and adequate soil drainage and temperature are necessary for their occurrence in an area. Diameters of tunnels range from four to eight inches, and tunnels are one to four feet below the surface. The tunnel system of an individual mountain beaver can extend three hundred feet in diameter. Several different types of chambers are constructed within a burrow system and are used for feeding, nesting, toilet, and refuse collection. Numerous other mammalian species have been documented using mountain beaver burrows, possibly as easy travel lanes. Average litter size is two to four altricial young, born in the spring following a gestation of about a month. Young are weaned at about two months of age. Except for dispersing individuals, mountain beavers rarely travel far from their burrow system. As such, park visitors are much more likely to see signs of mountain beaver activity than the animals themselves.

CONSERVATION CONCERNS

Mountain beavers are not hunted for sport or trapped for their fur. Because of the damage they do to new seedlings, mountain beavers are serious forest pests and are trapped to reduce their numbers in commercial forests. They negatively affect regeneration programs because they uproot or clip planted seedlings. Economic damage can be significant, and control programs involve direct removal of mountain beavers through trapping and toxicants, as well as nonlethal attempts to protect seedlings with physical barriers or chemical repellents.

MOUNTAIN BEAVER *Aplodontia rufa*	
COMMON NAMES	Boomer, Whistler, Sewellel
BODY LENGTH	13 inches (33 cm)
TAIL LENGTH	1 inch (2.5 cm)
BODY WEIGHT	2–3 pounds (0.9–1.4 kg)

BEAVER

Not only are beavers the largest rodents in North America (and the largest in the world after the capybara), no species has a greater impact on the environment other than humans. Likewise, no species ever figured so prominently in the early history and European exploration of the American West as did beavers.

North American beavers (there is a different species of beaver in Eurasia) occur throughout Alaska except for the northern tundra, all of Canada, and most of the United States except for southern Florida and arid portions of the southwest. The handiwork of beavers—downed trees, dams, and lodges—is easily seen in many national parks. The chunky appearance of beavers also is distinctive. However, because they are generally nocturnal and size is difficult to determine in the water, they could be mistaken for the much smaller muskrat.

The best-known physical feature of the beaver is the large, scaly, oval flattened tail, which serves a variety of purposes, although carrying mud (a popular misconception) is not one of them. The tail is used as a rudder while swimming, for thermoregulation, and to store fat. It also functions as a prop when a beaver sits on its haunches gnawing a tree, or as it works sticks or mud with the forelegs. The tail also serves to communicate danger when a startled beaver slaps it on the water. Following a warning slap, other beavers quickly swim to deep water, where they can remain submerged for up to fifteen minutes. Interestingly, tail slaps from young beavers are often ignored.

Less obvious than the tail are several other physical characteristics that also adapt the beaver to its aquatic environment. The eyes are high on the head so it can see while swimming on the surface, and there is a special sheath (nictitating membrane) that covers the eyes to protect and improve vision underwater. When submerged, the ears and nose are valvular and can be closed. Also, the lips close behind the large incisors to permit gnawing underwater. The hind feet are webbed for efficient propulsion while swimming, and the second claw is split as an aid in grooming. The skull is heavy with massive jaw muscles to accommodate gnawing through large trees. Incisors are kept sharp and beveled because the anterior surfaces are harder than the posterior side. As the beaver gnaws, the posterior sides wear down more quickly, producing the chisel-like cutting edges. Although steadily worn down, the incisors continue to grow throughout the life of an individual. Dense, thick outer pelage, usually a dark brown, overlays the heavy underfur and keeps beavers dry and warm.

Like muskrats, beavers are common along riparian areas: streams, ponds, lakes, or marshes where there is sufficient woody material for forage and constructing lodges and dams. Bark, leaves, and twigs of early successional species such as aspen, willow, poplar, and birch are commonly taken both for food and as building material. Large trees are gnawed all around, resulting in a characteristic cone-shaped cut. Small diameter trees are felled within a few minutes, although not always in the direction of the water (another popular misconception). Beavers certainly are "busy," however, and invest a large amount of time and energy in construction of dams and nests (lodges), as they cut trees and haul and place them. To build a dam, logs, vegetation, stones, and sticks are laid across a stream and secured with mud. Dams vary from a few feet high and ten feet long to huge structures ten feet high and a half mile or more in length. The goal is to create a pond with water deep enough not to freeze to the bottom during winter. The same woody building material is used for the lodge, which may be six feet high and twenty-five feet across. There is an insulated nest chamber above the waterline that is reached from underwater by one or more tunnels. Mud placed on the outside of the lodge freezes in winter, further securing the structure. On larger streams and rivers, beavers may not build a lodge but instead construct a den in the bank. Beavers in a colony mark the territory around their lodge or bank den by scent (castoreum) deposited from two pairs of glands at the base of the tail. The musky oil is placed on piles of mud or sticks specially constructed to serve as scent posts. The same woody material used in dams and lodges serves yet another purpose. It is cached in piles in the water near the lodge to provide a winter food supply. Stems are selected based on energy and nutrient content, as well as the amount of natural toxic compounds they contain. Stuck in the mud on the bottom of the pond, branches are reached by swimming under the ice, so beavers are not exposed to predators and the limbs are not frozen into the ice. Depending on the area, predators may include wolves, lynx, coyotes, mountain lions, fishers, and bears, but predation pressure has little impact on populations.

Depending on latitude, breeding extends from

(Pages 134–35) The largest rodent in North America, the beaver can remain submerged for up to fifteen minutes. Its thick fur keeps it warm in the coldest water, and it has excellent underwater vision.

Beavers have a greater impact on the landscape than any animal other than humans. A lunch of willow branches is provided by trees that grow adjacent to ponds created by beaver dams.

mid-January through February. Adult females first reproduce when three or four years old. Following a gestation of about 3.5 months, three to four young (kits) are born in May or June. Newborn kits are about an inch long and precocial, fully furred with their eyes open. They can swim when only a week old. Kits are weaned by three months of age but stay with the parents until the birth of the next litter the following year, when young adults are forced to disperse to new territories.

CONSERVATION CONCERNS

Beavers were trapped in North America for their fur as early as the 1600s. They were the driving economic motivation for the fur trappers or "mountain men" to explore the West in the early 1800s. Major fur-trading companies grew up around the trade in beaver pelts for the European market, and men like John Jacob Astor made fortunes. Unfortunately for beavers, their dams and lodges make them very easy to locate and trap. Without restrictions on the number of individuals trapped or season taken, populations were soon extirpated by overharvest throughout much of their historical range. Extensive logging throughout the 1800s also played a part in their extirpation. Recovery of beavers was due to wildlife management that limited seasonal harvest, as well as reintroductions during the mid-1900s to many areas throughout the natural range. Today, beavers are more widely distributed than they were historically. Beaver pelts still enter the fur market, but in many areas populations are considered to be pests because they cause flooding and damage to woodlands. Ecologically, beaver dams and the impact of resulting hydrological changes on the landscape are tremendous. The impounded water of beaver ponds affects all flora and fauna in a community, and provides habitat for aquatic invertebrates, fishes, waterfowl, and a variety of mammalian species from shrews to moose. Beavers eventually leave an area because they exhaust the surrounding food supply. Because it is not maintained, the dam begins to disintegrate, the pond drains or fills with silt and eventually becomes a wet meadow. After many years, the meadow becomes an early successional forest that supports different populations of plants and animals.

| **BEAVER** | |
Castor canadensis	
COMMON NAMES	Canadian beaver, North American beaver, Flat tail
HEAD AND BODY LENGTH	30–36 inches (76–91 cm)
TAIL LENGTH	9–13 inches (23–33 cm)
BODY WEIGHT	26–60 pounds (12–27 kg) average 100 pounds (45 kg) maximum

MUSKRATS

The largest of the mice and voles, muskrats are familiar to many people. Common to wet, marshy areas, streams, lakes, or farm ponds, muskrats look like small beavers but without the broad, flat tail. They have several physical features that allow them to spend a great portion of their lives in fresh water. The hind feet are large and partially webbed, with a fringe of stiff hairs between the toes (fimbriation) for additional surface area that provides propulsion. The tail is flattened along each side and acts as a rudder when the animal swims. The valvular mouth closes behind the front teeth (incisors), allowing muskrats to eat or carry food underwater. They have a dense pelage with long guard hairs and thick underfur to help them survive in often-frigid water. Color varies but usually is a dark brown to black with a lighter belly. Found only in North America, muskrats are widely distributed throughout Alaska south of the Brooks Range, almost all of Canada, and most of the United States except for Florida and parts of the arid southwest and Pacific Coast.

Muskrats can occur anywhere there is shallow water with abundant submerged and emergent aquatic vegetation. They opportunistically consume leaves, stems, and other parts of cattails, sedges, bulrushes, and numerous other plants depending on location, as well as crayfish, mollusks, and small vertebrates. They are generally nocturnal but also may be active during the day. Although common in many national parks, muskrats may not always be visible. Their workings usually are more evident, however. Muskrats construct their dens either in the banks of streams and rivers or in stick houses in marshes. Rounded houses, which contain the main nest, are three or four feet high—well above water level—and eight feet wide. They are accessed from under water and provide protection from predators and inclement winter weather. Winter can pose serious problems for muskrats because they are small and don't hibernate or store food. To remain warm and save energy during winter, several muskrats may huddle together in a warm, dry nest. During other seasons, however, muskrats are fairly asocial. Associated with the primary house are smaller "feeder huts" or "pushups." These also have tunnels through vegetation and

Shallow water and abundant aquatic vegetation are ideal for populations of muskrat, which are widespread throughout North America. Millions of muskrat pelts are taken annually for the fur industry.

underwater entrances that allow individuals to forage without being exposed to predators. Muskrats can swim underwater for close to two hundred feet before emerging to breathe. Mink are the primary predator on muskrats, but raccoons, coyotes, foxes, and various raptors also prey on them.

The "musk" in the common name muskrat refers to the odor produced from paired glands at the base of the tail in both sexes, especially evident during the breeding season. In northern areas breeding begins when the ice leaves ponds in the spring, and extends until late summer. Breeding may be year round in southern areas. Gestation is about thirty days. Mature females have two or three litters per year with an average of five or six altricial young (kits). Neonates are only about four inches long and weigh 0.8 ounce. Kits are able to swim before they are two weeks old and are weaned at about four weeks of age. Populations undergo a "fall shuffle" as young disperse to new areas in autumn. They are particularly vulnerable to predators at this time, especially if individuals must move through poor habitats. Average life span in the wild is probably about three years.

Another species, the round-tailed muskrat, fills the muskrat's niche in Florida and southeast Georgia. Round-tailed muskrats look like small muskrats but are about half the size and are more terrestrial with fewer aquatic adaptations. Their hind feet are not webbed and the tail is not flattened laterally. Like muskrats, however, they inhabit marshes, bogs, and lake shores where they feed on cane, sedges, rushes, and other wetland vegetation. Round-tailed muskrats also build dome-shaped houses with a nest chamber. Typically, these are smaller than those of muskrats, usually about two feet in diameter. A network of trails connects the nest house to a series of feeding platforms five to six inches in diameter. These provide protection while feeding, from predators such as snakes, raptors, and bobcats. Round-tailed muskrats breed throughout the year, with peaks in fall and winter. A female has four to six litters annually, with two to three young per litter. Gestation is twenty-six to twenty-nine days and newborns weigh only 0.4 ounce. Weaned by three weeks of age, round-tailed muskrats are sexually mature only nine weeks later.

CONSERVATION CONCERNS

Muskrats vie with raccoons as the most important furbearer in North America in terms of the number trapped each year. Millions of muskrats, worth tens of millions of dollars, are harvested annually in the United States and Canada during the winter when fur is thickest, longest, and most luxurious ("prime"). Interestingly, muskrat meat is occasionally served in some restaurants as "marsh rabbit." State and federal management agencies, as well as private landowners, manage marsh habitats and water levels to maintain population densities of muskrats for trappers. Census methods for muskrat populations involve both aerial surveys to count the number of houses visible and trends in harvest numbers relative to trapper effort. Conversely, during other times of the year muskrats can be serious pests when they disrupt farm irrigation dikes or their tunneling erodes river or stream banks. Population sizes are highly variable and fluctuate depending on winter weather, water conditions, predator pressure, food resources, parasites and disease, and other factors. Harvest is an important element in management because if muskrat populations in marshes become too numerous, the animals will overbrowse and seriously degrade the vegetation. These "eat-outs" adversely affect waterfowl populations and the muskrats themselves, and subsequent habitat restoration is often necessary. The round-tailed muskrat is not a significant part of the fur industry and, apart from occasional damage to dikes or agricultural areas, is of no economic importance. Populations have declined in some areas because development of canals has caused intrusion of saltwater into their habitat.

MUSKRAT	
Ondatra zibethicus	
COMMON NAMES	Mudcat, Muskbeaver
HEAD AND BODY LENGTH	9–12 inches (23–30 cm)
TAIL LENGTH	8–12 inches (20–30 cm)
BODY WEIGHT	1.5–4 pounds (0.7–1.8 kg)

ROUND-TAILED MUSKRAT	
Neofiber alleni	
COMMON NAMES	Florida water rat
HEAD AND BODY LENGTH	7–8 inches (18–20 cm)
TAIL LENGTH	4–7 inches (10–18 cm)
BODY WEIGHT	7–12 ounces (198–340 g)

PORCUPINE

Few species in the national parks or throughout North America are as easily recognized as the porcupine. This large, somewhat cumbersome rodent is the only mammal in Canada or the United States with quills. Quills are stiff, hollow modified guard hairs that cover much of the head, back, and tail. They may be five inches long on the rump, and there are up to thirty thousand quills on a large individual. Quills actually serve a variety of behavioral functions, but their primary purpose is protection against predators. Contrary to popular belief, porcupines do not intentionally throw quills at attackers. They can be erected and easily come loose, however, especially as an animal thrashes its tail when threatened or alarmed. Once lodged in the flesh of an unfortunate attacker, the quills are painful and difficult to remove. Microscopic barbs on the tips cause them to penetrate even further. The pelage is usually black or yellowish brown, and the quills are white with black tips. Although any of the larger predators can take porcupines, including wolves, coyotes, cougars, and bobcats, their primary predators are fishers. These members of the weasel family are particularly adept at turning a porcupine over on its back and attacking the belly, which is unprotected by quills. Unfortunately, however, people probably kill the greatest number of porcupines through shooting, poisoning, and accidents with vehicles.

Rodents are generally small, but porcupines are second only to beavers in body size among North American rodents. Males are slightly larger than females, and a very large porcupine can reach thirty pounds. Weight is variable, however, and during winter, individuals may lose 20 to 30 percent of their body weight.

Despite being slow-moving animals with poor eyesight and often described as lethargic, porcupines enjoy a wide geographic distribution. They occur from sea level to high elevations throughout most of Alaska, Canada, and the western Unites States, as well as the northeastern states. Porcupines are primarily forest dwellers but also are found in deserts, tundra, and grasslands. Their habitat variability is reflected in their diverse feeding habits. A great deal of conifer bark is eaten in the winter, especially pines and firs, but numerous deciduous trees are used as well. In warmer months, porcupines spend more time on the ground as they shift to eating grasses, flowers, and other succulent vegetation. Porcupines are active at night. Most likely to be seen around campsites in parks, they can be heard making soft grunting noises. Generally solitary, several individuals may den together in a log, rock crevice, or other natural cavity during the winter.

Breeding occurs in the fall and is a noisy affair. Males become vocal and highly aggressive, fighting each other with their sharp incisors and quills as they seek out receptive females. Females first reproduce when they are a year old; males are sexually mature at two to three years of age. Gestation

is very long for a rodent, about seven months, after which a single young is born. Neonates weigh about a pound and are precocial, moving about soon after birth with their eyes open. Quills on neonates are well developed, but are soft and pliable until about an hour after birth. Newborns have teeth and take vegetation by two weeks of age, although they continue nursing until four months old. Porcupines often live ten years in the wild, and rare individuals may survive twenty-five to thirty years.

CONSERVATION CONCERNS

Porcupines are fascinating, inoffensive mammals found in national parks throughout the country, but unfortunately in many areas their gnawing habits create serious damage. They are a concern for homeowners as they chew on wooden buildings, fenceposts, electrical wiring, and even vehicle tires. They also can cause significant damage to agricultural crops, destroy seedlings, and girdle mature trees. As a result, many areas have instituted control programs. Population densities are variable but remain secure in most regions. Generally, porcupines are a minor part of most wildlife conservation programs.

The porcupine feeds on a vast array of vegetation, including these pussy willows. With up to thirty thousand quills over its body, the porcupine is easily recognized. Among North American rodents, it is second in body size only to the beaver.

PORCUPINE *Erethizon dorsatum*	
HEAD AND BODY LENGTH	25–31 inches (64–79 cm)
TAIL LENGTH	6–12 inches (15–30 cm)
BODY WEIGHT	13–20 pounds (6–9 kg)

The sharply contrasting black-and-white color pattern and gray "saddle" behind its huge dorsal fin make the killer whale one of the most easily recognized cetaceans in the world. Killer whales are effective predators, as are all the toothed whales.

WHALES

Whales are some of the most fascinating, awe-inspiring, and least understood animals in the world. Carnivorous marine mammals that never leave the water, whales dive the deepest, make the longest migrations, and include the largest animals on earth. Whales (in the order Cetacea, from the Greek word for whale) can get so much larger than terrestrial mammals because the buoyant marine environment supports their massive body weight. They have streamlined or torpedo-shaped (fusiform) bodies that allow them to move through the water with minimal turbulence and resistance. Forelimbs are modified to form flippers and there are no external hind limbs. Power for swimming is provided by a horizontal flattened tail called the fluke.

There are two well-defined groups of cetaceans based on structural features: toothed whales (the suborder Odontoceti) and baleen whales (the suborder Mysticeti). Odontocetes have large, conical teeth that they use to forage on squid and fishes. Depending on the species, the number of teeth varies from two to more than two hundred. Instead of teeth, mysticetes have sheets of baleen that are used to strain small organisms from the water.

Toothed and baleen whales differ in several other physical and behavioral features as well. A toothed whale has a single external nare (nostril), often left of center on top of the head, a large forehead containing a fatty "melon," an asymmetrical skull, and complex nasal passages. Many of these physical features relate to the fact that the toothed whale echolocates to find prey. A baleen whale does not echolocate (although it produces a variety of audible sounds that travel great distances), has a symmetrical skull with two external nares, simple nasal passages, and no facial melon. Of the seventy-eight species of whales throughout the world, approximately thirty-nine are known or expected to occur in waters of the national parks; additional species no doubt remain unreported. We discuss six representative species because they occur in several of the national parks (see Table, pages 206–7) and are among the most likely to be seen. Whales often are very difficult to observe or identify, however. Sometimes a glimpse of a small portion of their back, dorsal fin, or fluke is all that is seen as these behemoths glide beneath the waves.

CONSERVATION CONCERNS

Whether small toothed whales or larger baleen whales, harvest of near shore, coastal species by natives throughout the world occurred for thousands of years as people used the meat, oil, baleen, and other byproducts for subsistence. These primitive operations involved open boats, nets, and handheld harpoons and had minimal effect on populations. Whale populations began to be affected by the 1500s and 1600s, as most European countries operated large ships throughout the world for the commercial harvest of whales. By the 1800s, the ever-increasing size of commercial whaling ships and efficiency of killing caused stocks of most species of whales to be overharvested. The depletion in whales continued in the early 1900s with the advent of floating factory ships. Many of the larger species were so depleted it was no longer economically practical to hunt them—they had become "commercially extinct." The International Whaling Commission (IWC) was established in 1946 to set quotas or prohibit harvest because most large species were on the verge of biological extinction. The IWC has no enforcement powers, however, and some countries continued to take whales for short-term economic gain. Today, the combined pressure of public opinion, government agencies, and numerous conservation groups helps limit whaling and promotes protection of whales and their habitat. Additionally, availability of alternative products lessens the demand for whale products and the economic incentive to hunt them. Most species of large whales are considered endangered or vulnerable today and are protected; only occasionally are a few taken by subsistence hunters. Smaller species such as Minke whales and Dall's porpoises continue to be harvested by certain countries, sometimes under the guise of "scientific collecting." However, populations of some larger species, such as humpback and blue whales, appear to be recovering well.

TOOTHED WHALES

Sperm whales are the largest odontocetes, with the smaller dolphins and porpoises making up much of the group. But even small whales are large compared to most other species of mammals. There are sixty-seven species of toothed whales that share several characteristics, including the fact that they echolocate. They send out low-frequency sound pulses that allow individuals to locate potential prey and orient in the water, based on information from the returning echoes. Toothed whales are generally smaller than baleen whales because they need to be fast and maneuverable to catch fish, squid, and other larger prey. Odontocetes also dive much deeper than ba-

leen whales to find prey. Sperm whales are the deepest divers, and have been documented at depths of over seven thousand feet, where water pressure exceeds 3,300 pounds per square inch.

Killer Whale

With its distinctive black-and-white color pattern and large, erect dorsal fin, the killer whale is one of the best known and most recognizable cetaceans. The heavy powerful body is black above and white below, with white oval patches behind the eyes and on the flanks. A gray "saddle" pattern is directly behind the dorsal fin. In the male, the dorsal fin is straight and up to six feet long; it is curved and a little over two feet long in the female. The killer whale's flippers are large, broad, and rounded.

The largest members of the dolphin family (Delphinidae), killer whales—also called orcas—are one of the most broadly distributed of all mammalian species. They occur throughout all oceans from the Arctic to the Antarctic regardless of water temperature or depth. Although solitary males may occur, killer whales usually form small groups (pods) of up to ten individuals. These "wolves of the sea" may travel a hundred miles a day as they prey on a variety of species. Marine mammals taken include seals, sea lions, sea otters, dolphins, porpoises, manatees, sperm whales, and even large baleen whales. Birds, turtles, fishes, squid, and octopus are also consumed. Like humpback whales, killer whales are highly acrobatic. They jump out of the water (breach), and

KILLER WHALE *Orcinus orca*	
MAXIMUM BODY LENGTH	30 feet (9 m) males
	25 feet (7.7 m) females
BODY WEIGHT	6 tons (5,450 kg) males
	4.2 tons (3,800 kg) females
DALL'S PORPOISE *Phocoenoides dalli*	
MEAN BODY LENGTH	7.5 feet (2.3 m) males
	7.0 feet (2.1 m) females
MEAN BODY WEIGHT	420 pounds (191 kg) males
	400 pounds (182 kg) females
HARBOR PORPOISE *Phocoena phocoena*	
MAXIMUM BODY LENGTH	5 feet (1.5 m) males
	5.5 feet (1.7 m) females
BODY WEIGHT	135 pounds (61 kg) males
	170 pounds (76 kg) females

may smash either the fluke (called lobtailing) or a flipper against the surface (called flippering). They also "spyhop"—raising the head vertically out of the water and rotating to "look around." Killer whales mate any time of the year. Gestation may be as long as eighteen months, and with a lactation period up to two years an adult female produces a calf only about every five years. Because they are top predators, killer whales are particularly vulnerable to accumulation of toxic contaminants. Small numbers are harvested in Asia, Greenland, and the West Indies, but populations generally are high.

Dall's Porpoise

One of six species in the porpoise family (*Phocoenidae*), Dall's porpoise is easily recognized by its very stocky, thickset body with disproportionately small head, flippers, and fluke. Because body color is black with a large white patch around the sides and belly, it can be mistaken for a killer whale calf. The small triangular dorsal fin is black and white, and the black flippers and fluke also are edged in white.

Dall's porpoises are found in cold coastal and deeper waters of the North Pacific. Well known for a plume of spray ("rooster tail") they create when surfacing, they are extremely fast, erratic swimmers and often ride the bow or stern waves of ships. They form small groups of a few animals, although hundreds may feed in the same area, taking squid and small schooling fishes. Interestingly, the teeth of Dall's porpoises are tiny, just a fraction of an inch long. Following a gestation of ten to eleven months, females have a single calf in June or July and breed again about a month later. Calves are three feet long and weigh about twenty-five pounds, and nurse for at least several months. Unlike most whales, Dall's porpoise may breed annually instead of every two or three years. They are preyed on by killer whales, and unknown numbers die each year when caught in fishing gear. Throughout most of their range, however, populations appear to be large and secure.

Harbor Porpoise

Besides Dall's porpoise, this is the only other species of porpoise in North American waters. Dark gray on the dorsal surface and lighter on the sides and belly,

harbor porpoises have a small triangular dorsal fin in the middle of the back. A dark line often runs from the corner of the mouth to the flipper. These small, unobtrusive toothed whales inhabit colder, shallow bays and harbors of coastal waters in the northern Pacific south to southern California, and in the Atlantic Ocean south to the Carolinas. Solitary or in small groups, they forage over long distances daily to prey on small squid, octopus, and schooling fishes like herring and mackerel (harbor porpoises are called "herring hogs" or "puffing pigs" in New England). Most calves are dropped in May, following a ten- to eleven-month gestation, and adult females mate again a month or two later. Calves may nurse for up to nine months, although they also begin taking prey before then. Throughout parts of their range, harbor porpoises are considered vulnerable because large numbers are killed annually in fishing gear. Large sharks and killer whales also take them.

BALEEN WHALES

With eleven species currently recognized, baleen whales (or mysticetes) represent only about 14 percent of living cetaceans. Although the number of species is small, mysticetes encompass most of the large whales, including fin, sei, right, humpback, and blue whales—the largest animals that have ever lived. Interestingly, within each species, females often are longer and heavier than males. Baleen whales are also known as rorquals (Norwegian for "tube-throated") for the series of parallel grooves or pleats on the throat. These allow the throat to expand during feeding as individuals skim or gulp large amounts of water. Water that enters the mouth, up to sixteen thousand gallons at a time in a blue whale, is not swallowed but with the mouth closed is forced outward through the baleen, which acts as a filter. Small marine organisms (collectively called zooplankton) and small schooling fishes are strained out and swallowed—several tons a day in larger whales. Baleen plates, which differ in size and shape depending on species, hang down from the upper jaw; there is no baleen in the lower jaw. Baleen is made of the same modified protein material (called keratin) that makes up fingernails in people. Because baleen whales feed on large concentrations of zooplankton floating in the water, they do not need to echolocate as toothed whales do. Also, baleen whales do not dive very deep because concentrations of their prey occur in relatively shallow water.

Gray Whale

True to their common name, the gray whale is a uniform mottled gray color. The entire body, but especially the head, is covered with white or yellowish blotches of barnacles and small host-specific crustaceans called "whale lice." Gray whales have no dorsal fin, but there is a series of bumps or knobs (called "knuckles") from the middle of the back toward the flukes. Whereas other baleen whales skim or gulp huge quantities of water to filter out prey, gray whales are the only mysticetes that are bottom feeders. In fairly shallow waters, they suck up muddy sediment containing large amounts of crustaceans and worms, which are then filtered through the short, coarse baleen. They feed during the summer in cold northern waters and then, like humpbacks, migrate up to five thousand miles south to shallow bays and lagoons in Baja California. There, in November and December, fe-

male gray whales have their calves and mate. Calves are dropped in January following a thirteen-month gestation. They weigh eleven hundred pounds at birth and are about sixteen feet long. They follow their mother north to the feeding grounds, where they are weaned at eight months of age. Gray whales once occurred throughout the northern hemisphere, but the population in the Atlantic became extinct within the last few hundred years for unknown reasons. Two distinct population segments remain in the Pacific. The eastern stock in North America extends from the Bering Sea south to Baja California. After being heavily overexploited, this population has completely recovered and was removed from the Endangered Species List in 1994. Because they migrate so close to shore, gray whales are easily observed and a mainstay of the whale-watching industry. Conversely, the western stock of gray whales in Asian waters is critically endangered with possibly only one hundred animals remaining.

Encrusted barnacles and "whale lice" adorn the head of a gray whale. These baleen whales are highly migratory and easily seen in shallow coastal waters.

The exceptionally long flippers of a humpback whale, evident as the animal breaches, give rise to the genus name meaning "big wing." Humpbacks are distributed worldwide and are a favorite of whale watchers.

Humpback Whale

One of the best-known large whales, the humpback is commonly observed and easily recognized because of its exceptionally long flippers. These are longer than in any other whale and equal one-third the length of the body (the genus name is derived from the Greek for "big wing"). Humpbacks are black above and usually lighter on the ventral surface. There are numerous small bumps (called tubercles) on the head and lower jaw, each with a single vibrissae (whisker). The posterior edge of the tail fluke is ragged, and the underside has a black and white pattern distinct to each individual. Humpback whales occur in oceans worldwide and are highly migratory. They feed throughout much of the year in cold, high latitude waters, and migrate to warmer tropical waters in the winter to calve and mate. Round-trip migrations may total ten thousand miles annually, matched only by those of gray whales. Calves, which are twelve feet long and weigh fifteen hundred pounds at birth, are born from January through March after an eleven-month gestation. Humpbacks are best known for their sensational acrobatic displays, including breaching, lobtailing, and flippering, although the functions of these behaviors are not clear. Another intriguing behavior of humpbacks is the eerie, mournful, highly complex songs that males sing during the breeding season. Males in a population sing the same song, often for days at a time, but the song gradually shifts through time. Within a year or two, a population has completely changed the structure and content of its song. Humpback whales were harvested commercially throughout the 1900s and are currently listed as endangered. Populations, however, appear to be recovering rapidly. Humpbacks, like gray whales, are a central focus of the whale-watching industry.

Minke Whale

Among baleen whales, only the pygmy right whale is smaller than the Minke whale. It occurs from subarctic to tropical waters throughout the Northern Hemisphere, although a smaller subspecies occurs in the Southern Hemisphere, as does the Antarctic Minke whale, a separate species. North American populations occur in the Pacific Ocean from northern Alaska to Baja California, and in the Atlantic from the Davis Strait to the Gulf of Mexico. As Minke whales migrate north in the spring, they move through shallow coastal waters, but they are in deeper water further offshore during southerly migrations in winter. Minke whales are black or dark gray on the back and white on the belly. A pale gray V-shaped band (chevron) sometimes occurs behind the head. The head has a narrow, pointed rostrum (the species name is Latin for "pointed beak") with a sharp ridge down the middle, the dorsal fin is fairly tall and curved back (falcate), and the top of each flipper has a distinctive white band. Minke whales travel alone or in small groups, and there is evidence that populations segregate by age and sex. They feed on small fishes, krill, squid, and octopus, and are themselves preyed upon by killer whales. Breeding occurs throughout the year, with peak calving depending on location. Gestation is about ten months, and calves nurse for four to five months. When large species were the focus of the whaling industry, Minkes were considered too small to harvest. They became important to whalers in the early 1900s as larger species became exceedingly rare due to overharvest. Norway continues to take Minke whales in the Atlantic and Japan does so in the Pacific, although international trade is banned. Population densities of Minke whales are believed to be secure.

GRAY WHALE *Eschrichtius robustus*	
MAXIMUM BODY LENGTH	49 feet (15 m)
MAXIMUM BODY WEIGHT	40 tons (36,300 kg)
HUMPBACK WHALE *Megaptera novaeangliae*	
MAXIMUM BODY LENGTH	56 feet (17 m)
MAXIMUM BODY WEIGHT	45 tons (40,900 kg)
MINKE WHALE *Balaenoptera acutorostrata*	
MAXIMUM BODY LENGTH	33 feet (10 m)
MAXIMUM BODY WEIGHT	10 tons (9,090 kg)

WEST INDIAN MANATEE

Large and easily recognized, manatees are docile grayish brown marine mammals. They are in the order Sirenia, a name derived from the sea nymphs of Greek mythology. When Columbus first saw manatees he believed they were mermaids. Although not closely related to whales, manatees share many of the same physical characteristics that adapt them to spend their entire life in water. Like whales, their only fur is short, stiff whiskers around the snout. They also do not have external ears, and their valve-like nostrils are on the top of the snout. Forelimbs are modified into paddle-like flippers, and there are no external hind limbs. The tail is distinctively round, flattened, and paddle-shaped. The dentition of manatees is very unusual. They have no incisors or canines, but only teeth in the back of their jaws (cheekteeth). Unlike other mammals in which new teeth erupt vertically from below, cheekteeth in manatees are replaced horizontally from the back of the jaw as existing teeth slowly move forward. This tooth replacement pattern, called mesial drift, also occurs in elephants, close relatives of manatees.

The West Indian manatee is one of three living species of manatee and the only one that occurs in North America. It ranges from Florida south through the Caribbean Sea, inhabiting shallow-water coastal systems of bays, estuaries, and inland river systems. Along with the closely related dugong, manatees are the only mammalian marine herbivores (whales eat other animals). They feed on submerged and emergent vegetation and may take plants overhanging the shore. They are restricted to shallow waters because the plants they feed on need sunlight. West Indian manatees have been found as far north as Rhode Island during the summer, but the population is concentrated in Florida and the Gulf Coast. Manatees are restricted to warm water around 68°F because they have poor ability to regulate their body temperature. They quickly lose body heat despite their large size, and cannot survive in cold water.

Manatees are often solitary, or occur in pairs or small groups. The species is not particularly social and the most common group is a cow and her calf. However, during winter groups of several hundred manatees may congregate around warm water sites such as discharge areas of power plants or inland lagoons. Vocalizations include audible squeaks and grunts, often between a cow and her calf at night or in turbid water. Males and females are sexually mature at three to four years of age. Mating occurs underwater, gestation is about one year, and most calves are born in spring and summer. A single young is most common although twins have been documented, which is very unusual among marine mammals. Newborn manatees are four feet long and weigh about sixty-five pounds. A female produces a calf only every three to five years.

CONSERVATION CONCERNS

The three species of manatees that exist in the world today are a remnant of the numerous species that existed worldwide twenty million years ago. The West Indian manatee is a federally endangered species and has been protected in Florida since 1893. Unfortunately, the species occupies an area

with ever-increasing numbers of people. Although individual manatees may live up to sixty years in the wild, human-associated mortality occurs through vandalism. Manatees also drown when they get pinned under flood control gates. The most significant cause of death for manatees, however, is being hit by boat propellers. Florida enacted regulations that restrict access to areas where manatees congregate, and in other areas reduce the speed of boats. Nonetheless, most adult manatees, if not killed, have scars along their backs from propeller accidents. Although manatees are very popular with the public, loss of habitat, entanglement in commercial fishing nets, and other developmental activities continue to negatively affect these slow, harmless, fascinating animals. Unfortunately, their long-term future remains tenuous.

The stocky, slow-moving West Indian manatee inhabits warm, shallow coastal waters where it feeds on submerged vegetation. Manatees are endangered, and several mortality factors reduce populations. In addition to habitat loss, one of the most significant causes of death is encounters with boat propellers.

West Indian Manatee	
Trichechus manatus	
COMMON NAMES	Manatee, Sea cow
LENGTH	7–12 ft (2.1–3.6 m)
BODY WEIGHT	330–790 pounds (150–359 kg)

CARNIVORES

Many of the best known and most charismatic mammals in North America and the national parks are in the order Carnivora. Carnivores encompass a broad array of species. National parks include the world's smallest carnivore—the least weasel—and one of the largest—the grizzly bear. Within the order, nine families are in national parks: the dog family Canidae (wolf, coyote, foxes); the cat family Felidae (cougar, lynx, bobcat); the bear family Ursidae (black and grizzly bears); the raccoon family Procyonidae (raccoon, white-nosed coati, ringtail); the weasel family Mustelidae (mink, weasels, wolverine, sea otter, river otter, marten, fisher, and badger); the skunk family Mephitidae; the sea lion family Otariidae (fur seals and sea lions); the earless seals, family Phocidae (elephant seals and others); and the walrus family Odobenidae. The last three families are marine carnivores often referred to as pinnipeds ("feather-footed") because their limbs are modified as flippers.

Despite their rich diversity in size and appearance, all carnivores share certain physical characteristics. The most obvious are their large canine teeth, which form the tusks in walruses. Many carnivores also have enlarged pairs of the last upper premolar and first lower molar. These carnassial teeth form a scissor-like shearing surface, and are especially well developed in canids and felids. Carnivores all have digits with claws, simple stomachs, and males have a penis bone (os baculum). Although the ordinal name would suggest members have a diet of meat, many species are omnivores. Raccoons, bears, and coyotes take a great deal of vegetation either seasonally or throughout the year. The sea otter takes invertebrates in the form of crustaceans and mollusks. The diet of cats, however, is entirely meat. Seals and sea lions consume primarily fish, and walruses eat clams, mussels, and occasionally fish and seals. Most carnivores are terrestrial, although the pinnipeds and sea otter spend most or all their lives in water.

Many carnivores are highly prized as furbearers because of their long, thick pelage. Historically, the marten and fisher provided high-value furs in North America, as did otters. Lynx have been trapped for hundreds of years, and of course mink—either wild or farm raised—continue to be economically important. Raccoons are one of the most popular game species taken by both trappers and hunters. Along with muskrats, raccoons make up about 70 percent of all the furs that enter the market in the United States. Subsistence hunting and trapping for all furbearers is allowed in Alaska's national parks except Glacier Bay and Kenai Fjords. Subsistence harvesting also occurs on those parts of Denali and Katmai that are national preserves. There are no seasons or bag limits in place for subsistence hunters.

A sad fact of history is that large carnivores are among the first species to be overly exploited or intentionally eradicated when people settle a new region. This was certainly the case as European settlers moved west throughout the 1700s and 1800s in the United States. Wolves were eliminated from most of their original range as were grizzly bears, and the range of the cougar was greatly reduced as well. Even some small carnivores, such as the black-footed ferret, remain highly endangered today. Fortunately, the dedicated efforts of conservationists and wildlife managers have allowed many species that were extirpated throughout much of their range to begin recovery—often within refuges and national parks.

EARED SEALS

All seals feed in the sea but give birth, nurse, and often mate on land or ice. Fur seals and sea lions are members of the family Otariidae. They are commonly called eared seals because of their small, cylindrical, external ears (pinnae), which do not occur in the hair seals (family Phocidae, also called earless seals). Eared seals also differ from hair seals in several other ways, including being able to rotate their hind flippers up under the body when on land. Thus, they move much more easily when "hauled out" than hair seals do. Also, fur seals and sea lions are only found in salt water, unlike phocids, which sometimes occur in fresh water. When in the water, eared seals derive most of their propulsion from their front flippers; hair seals primarily use the hind flippers. Sea lions have minimal underfur, and fur seals have thick, luxurious underfur; hair seals have no underfur. There are fourteen species of eared seals throughout the world, but only four species occur in North American waters: California sea lions, northern sea lions, northern fur seals, and Guadálupe fur seals. All are restricted to the Pacific Ocean and occur in at least one of the coastal national parks. Channel Islands National Park once hosted all four species, but northern sea lions have not been seen there since the 1980s.

Fur Seals

Northern fur seals have smaller heads and shorter, blunter snouts than Guadálupe fur seals. Also, the fur on the front flipper extends beyond the wrist in Guadálupe but not in northern fur seals, which also have longer hind flippers. Both species have uniformly dark, chocolate brown to black pelage, with females a lighter gray on the throat and chest. They have the typical seal body shape with heavy, robust forequarters tapering to slimmer hindquarters, with males significantly longer and heavier than females. Northern fur seals are distributed along coasts of the North Pacific Ocean and Bering Sea. Guadálupe fur seals are restricted to the coast of California and Baja California, with known pupping and breeding colonies only at Guadálupe and San Benito Islands. They are rarely seen in the Channel Islands.

Little is known about the life history of Guadálupe fur seals. Like northern fur seals, after the breeding season they spend extended periods from autumn through spring foraging at sea. Both species feed mainly at night, taking fish and small squid during fairly shallow dives (50 to 150 feet) lasting a few minutes. During hot summer weather, Guadálupe fur seals may be seen near shore and in tidal pools floating with a flipper in the air to help them stay cool.

Both species haul out on steep, rocky areas or sheltered caves near shore, often around tidal pools, to have their pups and breed. In North America, northern fur seals pup and breed in the Aleutian Islands of Alaska and the Channel Islands off California. Males establish territories in rookeries in May to attract females. Territories are defended by threat display, vocalizations, and fighting. Pups are dropped in June and July and mating occurs about a week after a female gives birth. Likewise, Guadálupe fur seals give birth and mate from May through July. Northern fur seal pups are about two feet long at birth and weigh ten to twelve pounds; Guadálupe fur seal pups are slightly smaller. Northern fur seal pups are weaned at about three months old, after which adult females quickly return to the sea. Female Guadálupe fur seals periodically nurse and forage for several days at a time for about nine months.

CONSERVATION CONCERNS

By the late 1800s, it was believed that Guadálupe fur seals had been hunted to extinction by com-

The small external ears and blunt snouts of northern fur seals are evident on mother and young.

GUADÁLUPE FUR SEAL	
Arctocephalus townsendi	
MEAN LENGTH	7.3 feet (2.2 m) males
	5 feet (1.5 m) females
MEAN BODY WEIGHT	418 pounds (190 kg) males
	110 pounds (50 kg) females
NORTHERN FUR SEAL	
Callorhinus ursinus	
MEAN LENGTH	6.6 feet (2 m) males
	4.6 feet (1.4 m) females
MEAN BODY WEIGHT	460 pounds (209 kg) males
	88 pounds (40 kg) females

Isolated rocky islets provide a safe refuge for sea lions and seals to haul out. These northern (or Steller) sea lions are substantially larger than the more familiar California sea lions.

mercial sealers. Apparently, however, a few dozen individuals remained at Guadálupe Island, which the Mexican government declared a sanctuary for seals in 1975. The population has since grown appreciably, although the species is still considered threatened. Interestingly, Guadálupe fur seals appear to have retained genetic diversity despite the extreme reduction in population size. Indigenous natives have hunted northern fur seals for hundreds of years, but populations only declined with the advent of commercial harvests in the 1700s and 1800s. The species was first protected from pelagic harvest, which primarily took females, by international treaty in 1911. Other regulations, including the U.S. Fur Seal Act and the Marine Mammal Protection Act, currently limit harvests, and the species numbers over a million individuals. Unfortunately, northern fur seals also die when they become entangled in huge sections of discarded fishing net floating in the Bering Sea and North Pacific Ocean, which the seals mistake for floating kelp beds and climb on.

Sea Lions

California sea lions are more familiar to people than any other species of seal in the world. They are easily trained, gregarious, and commonly seen in zoos and circuses. Large, vocal aggregations are frequently seen around marinas, docks, and buoys. Males have a uniform dark-brown pelage with a thick neck and torso, and a conspicuous forehead topped with lighter fur. Pelage of females and juveniles is a lighter tan or yellowish brown. California sea lions have long, slender snouts, whereas northern sea lions (also commonly called Steller sea lions) have larger heads and broader snouts. Pelage of northern sea lions varies from blond to tan dorsally and is darker ventrally, and males have a noticeable mane of long coarse hair draping the neck and shoulders. Northern sea lions are the largest eared seals and are considerably longer and heavier than California sea lions.

One subspecies of California sea lion is distributed along the Pacific coast from British Columbia to Baja California, with the largest breeding colony on San Miguel Island in Channel Islands National Park. Another subspecies, the Galápagos sea lion, occurs on the Galápagos Islands. The last subspecies, the Japanese sea lion, has probably been extinct since

CALIFORNIA SEA LION *Zalophus californianus*	
MEAN LENGTH	7 feet (2.1 m) males 5.3 feet (1.6 m) females
MEAN BODY WEIGHT	825 pounds (375 kg) males 207 pounds (94 kg) females
NORTHERN (STELLER) SEA LION *Eumetopias jubatus*	
MEAN LENGTH	9.2 feet (2.8 m) males 7.6 feet (2.3 m) females
MEAN BODY WEIGHT	1,245 pounds (566 kg) males 579 pounds (263 kg) females

the 1950s due to commercial overharvest. Northern sea lions are found along the North Pacific from northern Japan south to California, with breeding colonies from Alaska to southern California.

Sea lions feed opportunistically, with diets shifting by locality and season depending on available prey. Generally, they take a variety of fishes such as anchovies, sardines, cod, pollock, and mackerel, as well as octopus and small squid. California sea lions usually dive less than three hundred feet deep for a few minutes, although they can exceed seventeen hundred feet in depth. Likewise, most dives of northern sea lions are fairly shallow and of short duration. While at sea foraging for weeks or months at a time, sea lions dive almost continuously, spending little time at the surface. Near shore, California sea lions are extremely bold, as they may rob fish from commercial nets or fishermen's lines. Sea lions are preyed upon by larger species of sharks and by killer whales, but predation has little effect on population densities.

Like other seals, sea lions leave the water to have their pups and breed. They haul out on rocky ledges or reefs, tidal flats, or sandy beaches, with northern sea lions usually preferring more isolated areas than California sea lions. As in fur seals, adult bulls establish territories beginning in May to have access to breeding females. Males hold territories through aggressive threats, clamorous and incessant vocalizations and bellowing roars, and occasional fighting. Most females of both species drop their pup in late June and breed again two or three weeks later. Pups are nursed between periodic foraging bouts. Returning to shore, a mother recognizes her pup among all the others by individual voice and smell. By six months of age, most California sea lion pups are weaned and abandoned to fend for themselves, although some continue to nurse for a year or more. Northern sea lion pups are weaned when they are a year old.

CONSERVATION CONCERNS

Like most coastal species of seals, northern and California sea lions have been hunted by indigenous peoples for thousands of years. Throughout the early 1900s, a bounty was placed on both species because they were considered to compete with commercial fishermen. California sea lions continue to suffer mortality through entanglement in gill nets, but populations are stable, whereas the Galápagos sea lion is considered threatened. Since the 1970s, populations of northern sea lions have declined dramatically. Whether this is due to the effects of commercial fisheries on the prey base, natural long-term environmental changes in the ocean, diseases, or a combination of these and other factors is unknown. The western stock of northern sea lions is currently endangered, and the eastern stock is considered threatened.

HAIR SEALS

All seals feed in the sea but haul out on land or ice to give birth, nurse, and mate. Hair seals (family Phocidae) are also called true seals or earless seals. Several morphological and anatomical features separate them from the fur seals and sea lions (family Otariidae). The most easily observed are that phocids do not have the small cylindrical external ears of the fur seals, although they can hear well. Also, their flippers are fully furred with nails of equal size, and the hind flippers do not rotate up under the body. Thus, when hair seals are "hauled out" on land, they drag their bodies along rather awkwardly by means of the front flippers. Additionally, phocids have guard hair but no underfur, and the cheekteeth have numerous cusps. Phocids also dive deeper for prey and stay submerged longer than otariids but are less agile on land. They prey primarily on fishes and small marine animals. Like all seals, they exhibit delayed implantation of the fertilized egg. There are nineteen species of hair seals found around the world, primarily in polar or temperate waters. Only the endangered Hawaiian monk seal inhabits warm water. Caribbean monk seals have not been seen since the early 1950s and probably are extinct. Ten species of hair seals are in North American waters, but only three—the gray seal, harbor seal, and northern elephant seal—occur in national parks.

Harbor Seal

Medium-sized and active during the day, harbor seals are one of the most commonly seen and widely distributed of the hair seals. They occur throughout much of the Northern Hemisphere in both the Atlantic and Pacific Oceans, where they are commonly seen in shallow bays and estuaries, around piers and beaches, or hauled out on tidal flats or rocky islets. Torpedo-shaped with a large round head and broad snout, their flippers are fairly small. The short coarse pelage consists of light, uneven, asymmetric rings against a darker background (for individuals in southern portions of the range), or a silver background with darker-colored rings. Harbor seals in the Pacific are slightly heavier than those found in the Atlantic.

commercially important fish, bounty programs in the United States and Canada once killed tens of thousands of individuals annually. Native subsistence hunters continue to harvest harbor seals for food and other products. Today, however, populations on both the Atlantic and Pacific Coasts of North America are considered both secure and quite high.

Gray Seal

Gray seals occur throughout the northern Atlantic Ocean and Baltic Sea. In North America, they inhabit the coastlines of Labrador, Newfoundland, and Nova Scotia around the Gulf of St. Lawrence, and south to Nantucket Island, Massachusetts. Male gray seals are larger and heavier than females; those in Europe are smaller than seals in North America.

Pelage of males is black or a dark, mottled brown with lighter patches. Females usually are grayish with darker markings. Spotting patterns may be unique to each gray seal and can be used to identify individuals. Gray seals have large upper bodies that taper toward the hindquarters, and short heavy flippers. The distinctive head, especially in males, is large with a broad fleshy snout. An alternative common name is the horsehead seal (and the rather unflattering scientific name means "hook-nosed sea pig").

Gray seals are much more gregarious than harbor seals. Hundreds may congregate on rocky coastal areas, beaches, or ice floes. They feed on cephalopods and various species of fishes, often on the sea floor. Gray seals may remain at sea for three weeks while foraging.

Breeding occurs in January or early February as males fight each other for access to females. Successful males will mate with fewer than ten females, a limited number compared to many other species of seals. Female gray seals produce a single pup each year. Birth takes place either on land or on ice floes. Males weigh about thirty-eight pounds at birth, and female pups are about thirty-two pounds. Although nursing lasts only about sixteen days, the pup gains four to five pounds a day because the fat content of the milk is 40 to 60 percent. Conversely, a nursing female may lose up to thirteen pounds a day as she converts blubber to milk. When her pup is weaned, a female mates again, on land, on an ice floe, or in the water. Gray seals may live up to forty years in the wild.

The harbor seal has relatively short flippers and, like all hair seals, lacks small external ears. Widely distributed throughout shallow coastal waters of the Northern Hemisphere, the harbor seal is commonly seen in national parks on both the Pacific and Atlantic coasts.

Like other phocids, harbor seals feed on fish, cephalopods (octopus, squid, and cuttlefish, for example), and crustaceans. Diets vary depending on location and season of the year. Harbor seals can dive as deep as fourteen hundred feet and remain submerged for thirty minutes, but most dives are less than three hundred feet and last only a few minutes.

The breeding season depends on latitude, but adult females generally have a single pup each year. Both sexes reach sexual maturity between three and seven years of age. Mating occurs in the water and implantation of the fertilized egg is delayed for two to three months. Once implantation occurs, actual gestation is six to seven months. The precocial pups weigh about twenty-two pounds and can swim almost immediately after birth. Nursing lasts for about six weeks, and after her pup is weaned a female mates again. Prime predators include killer whales, sharks, and northern sea lions in the water, and coyotes and grizzly bears on land. Harbor seals also suffer mortality from diseases such as influenza and distemper. Although a harbor seal may live twenty-five years, most pups probably do not live long enough to breed.

CONSERVATION CONCERNS

Because harbor seals were believed to feed on

CONSERVATION CONCERNS

Gray seals have been harvested along the coasts by subsistence hunters for thousands of years. Because they take commercially valuable fish and damage fishery equipment, bounties were paid in Europe throughout the 1900s to reduce populations. Today, only a remnant population of a few thousand gray seals remains in the Baltic Sea area. The North American population is large and increasing, however. There are also conflicts between conservation groups and the fisheries industry in North America over the gray seal taking commercially valuable species. Additionally, gray seals carry a parasite, called sealworm, which unfortunately infects fish. The parasite is not harmful to people, but must be picked from fish fillets before they are sold, which is a very expensive process.

Northern Elephant Seal

These are the largest North American seals. Males are significantly larger than females, and can be four times heavier, reaching over four tons in body weight. Their large size, plump flexible snout (proboscis), and a thickened neck shield that is often scarred and callused distinguish males from other phocids. The body is heavy and robust in the forequarters, and thinner toward the hindquarters, and the flippers are fairly short. Pelage color is dark brown above and yellow below, with females and juveniles lighter in color than adult males. Northern elephant seals haul out on beaches from

Harbor Seal *Phoca vitulina*	
MEAN LENGTH	6 feet (1.8 m) males 5 feet (1.5 m) females
MEAN BODY WEIGHT	330 pounds (150 kg) males 264 pounds (120 kg) females
Gray Seal *Halichoerus grypus*	
MEAN LENGTH	7.6 feet (2.3 m) males 6.6 feet (2.0 m) females
MEAN BODY WEIGHT	627 pounds (285 kg) males 440 pounds (200 kg) females
Northern Elephant Seal *Mirounga angustirostris*	
MEAN LENGTH	12.5 feet (3.8 m) males 8 feet (2.4 m) females
MEAN BODY WEIGHT	3,960 pounds (1,800 kg) males 1,320 pounds (600 kg) females

Vancouver Island, British Columbia, south to Baja California, where they are fairly docile despite their large size.

Northern elephant seals are probably solitary throughout most of the year as they forage in the central and North Pacific Ocean as far north as the Aleutian Islands of Alaska. They are submerged most of the time while at sea as they dive for squid and fishes. Because of their large size, they are able to descend over five thousand feet deep and stay down for two hours. Most dives, however, are probably around two thousand feet deep and last about twenty minutes.

Females come ashore from December through February on sandy or rocky beaches to give birth and breed. Adult bulls are large enough by about eight years of age to establish dominance hierarchies to gain access to breeding females at the rookeries. They use vocal and visual threat displays toward other males, inflating the proboscis to help resonate vocalizations, which can be heard a mile away. Occasionally, these raucous threats escalate to actual fighting as combatants pummel and bite each other around the head and thick skin of the neck. A successful male may mate with up to fifty females. Implantation of the fertilized egg is delayed for two to three months. Once implantation occurs, actual gestation takes about nine months. Females have a single pup, which weighs about seventy-five pounds at birth. Females and pups remain on shore and nurse for about four weeks. By the time they are weaned, pups weigh over three hundred pounds and their mother has lost about 40 percent of her body weight. After mating, the female returns to the sea to feed, leaving her pup alone on shore. About a month later, pups enter the water and begin to forage on their own.

Adult bull northern elephant seals spar to establish dominance and access to females. As the animals pummel each other for territorial breeding rights, an inflated proboscis causes their roaring vocalizations to resonate.

CONSERVATION CONCERNS

Commercial sealers hunted northern elephant seals throughout the 1800s for their blubber, which was used to make oil for lamps and as a lubricant. By 1900, they were harvested almost to the point of extinction, with perhaps as few as two dozen individuals remaining at Guadalupe Island, Baja California.

Since then, the population has recovered to more than 150,000 seals, with the largest breeding colonies in and around the Channel Islands. There appears to be little genetic diversity, however, probably because of the population "bottleneck" resulting from extreme overharvest.

Coyotes are one of the most widespread and adaptable carnivores in North America. Unfortunately, they also are a controversial species, creating contentious debate as to whether they should be viewed as fascinating wildlife or a pest species.

COYOTE

Highly adaptable and opportunistic in their habitats, diets, and behaviors, no other mammalian species defies generalization as much as the coyote does. Likewise, few species generate as much controversy among competing interest groups. Coyotes are medium-sized canids, often said to resemble small German shepherds, with males slightly larger and heavier than females. A long, thin muzzle, long, erect ears, and smaller size help differentiate coyotes from wolves. Identification is sometimes difficult, however, because coyotes and wolves hybridize in some areas. Coyotes also breed with domestic dogs, although hybrid "coydogs" are uncommon. Pelage in the coyote (pronounced ky-YO-tee, after the Aztec word *coyotyl*) is long, coarse, and dense. Fur is dark gray or brownish above, with a darker band along the back, and somewhat lighter below. The neck and sides may have a reddish tint, and the tail is long and bushy with a black tip. When coyotes run, they hold their tail down or between the hind legs; wolves and dogs run with the tail held out or above the back.

Geographic distribution of coyotes is widespread throughout most of Alaska, Canada, and the United States, extending south into Central America. Once absent in the eastern United States, coyotes have increased their range and population density dramatically because of increased land cleared for agriculture, logging of forests, and extirpation of larger carnivores. Given their wide distribution (suggested by the number of national parks where they occur), it is not surprising that coyotes have adapted to an array of available habitats, including deserts, grasslands, wetlands, mountains, agricultural areas, and suburban developments. Feeding habits are as broad and diverse as the habitats occupied and vary seasonally as well as geographically. Coyotes are primarily carnivores. They commonly take rodents by stalking and pouncing, and chase down rabbits and hares. Coyotes also consume a variety of other vertebrates, invertebrates, eggs, vegetation, and fruit. Often blamed for preying on calves and lambs, as well as deer and elk, many of the domestic livestock and wild ungulates taken are scavenged as carrion. Coyotes are active throughout the year. Primarily nocturnal, they also may be seen foraging during summer days.

Group size is variable, although coyotes do not have as involved a social structure as wolves. Single animals or pairs are common, although groups of three to eight individuals may hunt together. Social dynamics are maintained and enhanced by highly developed visual cues and scent marking. Like domestic dogs, coyotes scent mark by urinating on stumps, rocks, and bushes to advertise the boundaries of their area. Coyotes also have a variety of vocalizations (the species name *latrans* is Latin for "barker"). They are well known for their characteristic high-pitched mournful howling, which occurs throughout the year. Howling serves to announce location or affirm individual identity. Occasionally, individuals or groups may howl simply because they enjoy it.

Dens are constructed in various places, including brushy hillsides, hollow logs, and rocky areas. A coyote may excavate its own burrow or simply enlarge abandoned fox or woodchuck diggings. Tunnels are about a foot in diameter and up to thirty feet long. Dens, used to raise young, are usually just bare dirt three feet in diameter. Male-female pair bonds may remain stable for several years, though not necessarily for life. Coyotes first breed when they are a year

Coyote *Canis latrans*	
COMMON NAMES	American jackal, Brush wolf, Prairie wolf
HEAD AND BODY LENGTH	2.6–3.2 feet (0.8–1.0 m)
TAIL LENGTH	12–16 inches (30–41 cm)
BODY WEIGHT	20–40 pounds (9–18 kg)

old and produce a single litter a year. Breeding begins in January or February. After a gestation of about sixty-three days, a mean litter of six pups (whelps) is born in April or May. Whelps are covered with a short, fine fur, but otherwise are altricial and weigh about nine ounces. They first emerge from the den at two to three weeks of age and are weaned about a month later. By autumn, young have dispersed from the natal area. Although wolves and mountain lions kill coyotes, most mortality is caused by humans.

CONSERVATION CONCERNS

Many people consider coyotes an abiding symbol of the Old West. Many farmers and ranchers, however, consider them "varmints" to be eliminated. Additionally, in suburban areas, coyotes are often responsible for preying on domestic pets such as cats and dogs. State and federal Animal Damage Control programs attempted to eradicate coyotes over many decades of the twentieth century, and tens of thousands were killed annually. At one time or another, almost every state has had a bounty on coyotes, none of which were effective at reducing populations. Control methods include shooting, trapping, and poisoning, as well as a variety of non-lethal methods. This wide-scale approach to eradication proved fruitless because of the ecological resiliency of the species. Today, control efforts in rural areas are focused on site-specific "problem animals" as opposed to the entire coyote population. On the positive side, coyotes reduce rodent and rabbit populations that otherwise could cause great economic loss. Coyotes also are important in the fur market, with about 500,000 harvested annually. Debate will continue over the relative merits of coyotes, and the array of biological, social, economic, legal, and ethical questions associated with efforts to control them.

GRAY WOLF

With their plaintive "call of the wild" howls, few mammalian species have the mystical charisma of gray wolves. Wolves and humans have a long history of interrelationship. Wolves were the first mammals to be tamed and domesticated, about twelve thousand years ago, eventually resulting in the hundreds of different breeds of dogs throughout the world today. In North America, wolves are distributed throughout Alaska and Canada. South of Canada, wolves occurred throughout the United States to central Mexico prior to European colonization but, like grizzly bears, were almost completely eradicated through state and federal predator control programs concurrent with settlement. Today, wolf populations are found in remote parts of northern Minnesota (including Voyageurs National Park), Wisconsin, and Michigan (including Isle Royale National Park), as well as Washington, Idaho, Montana (including Glacier National Park), and Wyoming (reintroduced to Yellowstone and Grand Teton National Parks). Wolves also were reintroduced in southern Arizona and New Mexico in 1998.

The gray wolf is the largest wild member of the dog family (Canidae) in the world. A large adult male may weigh 175 pounds, although average weights are much less. Males are somewhat larger than females, and individuals at northern latitudes are larger than those further south. The long, coarse pelage is usually gray to brown, darker on the back and tail, with white fur around the mouth. Pelage color can vary among individuals from white to black. The wolf is much larger and heavier than the coyote, and it has a broader snout and much larger feet. It runs with the tail held out, whereas the coyote runs with the tail nearly between the hind legs.

Wolves are habitat generalists within their large territories, although specific habitat types are used for foraging and selection of den sites depending on local conditions. Numerous factors affect habitat and den selection, including prey availability, latitude and prevailing weather conditions, and degree of human development, roads, and presence of livestock. Dens can be in any sheltered place, including rocky ridges, small caves, or excavated tunnels. Several dens may be constructed and used in proximity to the core use area.

Wolves are highly social. Although solitary individuals occur, wolves generally form packs of five to twelve individuals. Larger packs have been documented, however. A dominance hierarchy within each pack eliminates prolonged aggressive interactions. An alpha male and female lead a pack and do the breeding, and pairs stay together for years. Several related wolves are usually pack members, and unrelated individuals may sometimes occur. Communication among wolves in a pack is essential to maintain the dominance hierarchy; communication between packs is necessary to maintain and defend territories. Just as in domestic dogs, communication involves overt and more subtle visual displays, scent marking, and vocalizations. Howling is well known and functions in pack cohesion and delimiting territories. Size of territories varies, but can be as large as five thousand square miles, with individuals traveling as much as fifty miles a day. Wolves that follow migrating caribou herds travel

hundreds of miles back and forth seasonally.

True carnivores, wolves prey primarily on deer, caribou, elk, moose, mountain sheep, muskoxen, and other ungulates. Most large animals taken by wolves are young, old, or in poor health, and predation does not negatively affect healthy herds that have sufficient habitat. Regardless, an integrated, well-functioning pack is essential to stalk, chase, and kill large, potentially dangerous prey successfully. Wolves take a variety of smaller animals as well, such as hares, beavers, smaller carnivores, and rodents. Calves, sheep, and other livestock are also preyed upon. Lack of vigilance and slow flight makes domestic animals easy targets compared to wild species. Wolves also may scavenge carrion.

Wolves have significant impacts on ecosystems and competitively displace other carnivores. Densities and distribution of small to medium-sized carnivores such as coyotes, foxes, and wolverines, often called mesocarnivores, change as wolves kill or drive off these subordinate species. Wolves also have an impact on larger predators, with fatal interactions documented with bears and mountain lions. Wildlife biologists are now studying the extent of changes on mountain lion ecology related to the reintroduction of wolves in Yellowstone National Park. The predator-prey dynamics of wolves and moose on Isle Royale have been studied for decades. Populations are generally in equilibrium, but extrinsic factors such as weather can periodically disrupt the balance. Moose populations are significantly affected by winter snow conditions and resulting forage availability. The population density of wolves depends on that of moose, secondarily on the density of alternative prey such as beavers, and is also limited by the size of the island.

The breeding season extends from early winter in southerly latitudes to spring in more northern areas. Adult females are receptive for a week or two, during which pairs exhibit courtship behavior. Following a nine-week gestation, an average litter of six altricial pups, each weighing about one pound, is born in a natal den. Other pack members bring food to the mother and pups for the first three or four weeks. Pups are weaned about a month later and moved to a different "rendezvous" area, where they play, bond with each other, and become integrated into the social structure of the pack. Pups grow to nearly adult size by autumn and join the pack as it hunts. Within the next year or two, young leave the pack, dispersing to new areas sometimes hundreds of miles away. They become sexually mature at two years of age.

CONSERVATION CONCERNS

Few species have generated as much controversy

GRAY WOLF *Canis lupus*	
COMMON NAMES	Timber wolf, Tundra wolf, Prairie wolf
HEAD AND BODY LENGTH	43 inches (109 cm)
TAIL LENGTH	18 inches (46 cm)
BODY WEIGHT	110 pounds (50 kg) males 77 pounds (35 kg) females

and debate over conflicting management goals as wolves. Farmers and ranchers generally consider them livestock killers, but to many other people they excite the imagination as symbols of wild places. Wolves were extirpated from 98 percent of their former range in the lower forty-eight states through trapping, shooting, and poisoning as part of state and federal predator removal programs. Recently, renewed wolf control has been proposed in parts of Alaska, including Denali National Park, in an effort to bolster declining caribou populations. Conversely, wolves have been the focus of conservation and reintroduction programs south of Canada to establish populations in remote areas away from human interference. One such high-profile reintroduction was in Yellowstone National Park. The last wolf in Yellowstone was shot in 1926. After decades of debate between the U.S. Fish and Wildlife Service (USFWS), conservation groups, and ranchers, fourteen wolves from western Canada were released in Yellowstone in 1995, with an additional release in 1996. This reintroduction was very successful; density and distribution increased, and now fifteen thousand park visitors a year see a wolf in Yellowstone.

Since April 2003, the USFWS designated three gray wolf Distinct Population Segments in the contiguous forty-eight states. The eastern and western segments are listed as threatened, while the southwest segment is considered endangered. Additionally, reintroduced wolves in Yellowstone National Park and in Arizona and New Mexico are considered Nonessential Experimental Populations. This means that these geographically separated populations are not considered "essential" for the continued existence of the species. If they leave a refuge or a national park, they do not have the legal protection of a listed species. In July 2004, the USFWS considered the eastern population segment of gray wolves "recovered" and proposed removing them from the list of endangered species. Conservation groups disagreed and the debate continues.

FOXES

Shy and reclusive, foxes are the smallest members of the dog family (Canidae), which in North America includes the gray wolf and coyote. There are six species of foxes in the United States, and each occurs in at least one national park. By far the most easily recognized and widely distributed are red foxes and gray foxes.

Red Fox

One of the most widely distributed carnivores in the world, the red fox is found in Europe, Asia, and Africa, and was introduced into Australia in the mid-1800s for fox hunting. In the 1700s, red foxes from Europe were introduced into the native population in the United States as well. In North America, they range throughout Alaska, Canada, and the United States except in the arid southwest. With a fairly long pointed muzzle and large erect ears, red foxes are the size of a small dog. Males (called dogs) are slightly larger than females (vixens). The long dense pelage is reddish orange to rusty red above, with white cheeks, throat, and belly. Feet and lower legs are black, as are the back of the ears, and the long, bushy tail has a white tip. Variations in pelage color range from silver to black and give rise to other common names.

Red foxes are not deep-forest animals. Good red fox habitat is mixed grasslands, borders of woodlots, riparian sites, pastures, brushy thickets, and suburban areas around parks, golf courses, and cemeteries. They hunt and scavenge opportunistically, primarily at night. Small rodents and rabbits are a large part of the diet, as well as ground-nesting birds, eggs, fruits, and invertebrates. Red foxes have insignificant effects on game bird or mammal populations, but occasionally raid poultry or other farm animals. Dead ungulates and livestock are eaten as carrion, and excess food may be cached. Red foxes dig their own dens, or they may take over and enlarge those of marmots or badgers. They also den in hollow logs, caverns, and old abandoned buildings.

A mated pair shares a territory, and breeding can occur from January through March. Yearling females may breed depending on population density and food availability. Following a gestation of about fifty-two days, an average of five pups is born in the den. Pups weigh three to four ounces at birth and nurse for eight to ten weeks. The family breaks up as pups disperse in the autumn. The parents come together again the following breeding season.

CONSERVATION CONCERNS

During European settlement red fox populations in the United States probably increased in density and distribution due to extensive logging and the elimination of wolves and coyotes. Conversely, recent increases in coyotes have no doubt reduced the number of red foxes in many areas. Red foxes are economically valuable as furbearers. About thirty-eight states have a trapping season on them. Average annual harvests amount to nearly 200,000 animals, worth $2–3 million depending on pelt prices. Far greater amounts are spent to control populations, however, because red foxes are a significant vector in the spread of rabies. Control efforts have involved poisoning, trapping, gassing, and offering bounties in efforts to reduce red fox populations. Oral vaccines have also been used in attempts to vaccinate red foxes against rabies. Another mortality factor for red foxes is collisions with vehicles, and farm machinery also takes a toll. Nonetheless, populations are stable throughout their range.

Gray Fox

Unlike the red fox, the gray fox is closely associated with deciduous forests. It also occurs in brushy and grassland habitats throughout most of the United States, except for much of the Northwest. Its geographic range extends south through Mexico to northern South America. It is more shy and secretive than the red fox, and not as closely linked to developed areas. Slightly smaller than the red fox, and with shorter, coarser fur, the gray fox is easily recognized by its grizzled, salt-and-pepper silver-gray pelage on the dorsum (the species name combines two Latin words for "gray" and "silver"). There is a black band down the middle of the back extending to the black-tipped tail. The belly is a buff white, and the sides and legs are a rust color.

Active primarily at night throughout the year, gray foxes are omnivorous. They feed opportunistically on plants and animals, generally mice and voles, as well as rabbits and invertebrates. The diet shifts seasonally and corn, apples, and fruits are taken when available. Gray foxes are unique among North American canids in that they can climb fifty feet high or more into trees. They climb either to escape danger or to forage on fruits or squirrels, and are agile enough to jump from limb to limb. Like other foxes, they den in rocky outcrops, hollow logs, dense brush, and burrows but rarely construct their own dens.

In North America, reproductive activity is similar to that of the red fox. Pairs mate for life, and breeding occurs from January through March. Average litter size is four pups born following a fifty-nine-day gestation. Newborn pups weigh only about three ounces but grow quickly and are able to hunt when four months old. Young disperse in the fall, often fifty miles or more.

CONSERVATION CONCERNS
Like red foxes, gray foxes are trapped for their fur

in about forty states, with annual harvests generally around 100,000 animals. Gray foxes also carry rabies, and research has tested delivery systems for oral vaccines. Because they are more closely associated with forested habitats, the range of gray foxes has not expanded as much as that of red foxes. Regardless, gray fox populations are generally stable.

Other Foxes

As their name suggests, arctic foxes are circumpolar in arctic habitats in both the Eastern and Western Hemispheres. They occur on ice floes and coastal tundra in northern and western Alaska and have

ARCTIC FOX *Alopex lagopus*	
COMMON NAMES	Polar fox, White fox, Blue fox
HEAD AND BODY LENGTH	27 inches (69 cm)
TAIL LENGTH	12 inches (30 cm)
BODY WEIGHT	5.5–15 pounds (2.5–6.8 kg)
GRAY FOX *Urocyon cinereoargenteus*	
COMMON NAMES	Tree fox, Cat fox, Grayback
HEAD AND BODY LENGTH	24 inches (61 cm)
TAIL LENGTH	14 inches (36 cm)
BODY WEIGHT	6.6–15.4 pounds (3–7 kg)
ISLAND FOX *Urocyon littoralis*	
COMMON NAMES	Coast fox, Short-tailed fox
HEAD AND BODY LENGTH	19 inches (48 cm)
TAIL LENGTH	8 inches (20 cm)
BODY WEIGHT	2.2–6 pounds (1.0–2.7 kg)
KIT FOX *Vulpes macrotis*	
COMMON NAMES	Desert fox
HEAD AND BODY LENGTH	18 inches (46 cm)
TAIL LENGTH	11 inches (28 cm)
BODY WEIGHT	3.7–6.6 pounds (1.7–3.0 kg)
RED FOX *Vulpes vulpes*	
COMMON NAMES	Cross fox, Silver fox, Black fox
HEAD AND BODY LENGTH	23 inches (58 cm)
TAIL LENGTH	15 inches (38 cm)
BODY WEIGHT	6.6–17.6 pounds (3–8 kg)
SWIFT FOX *Vulpes velox*	
COMMON NAMES	Plains fox, Prairie fox
HEAD AND BODY LENGTH	19 inches (48 cm)
TAIL LENGTH	11 inches (28 cm)
BODY WEIGHT	4.2–5.3 pounds (1.9–2.4 kg)

been found within ninety miles of the North Pole. Arctic foxes feed on voles, lemmings, hares, ground-nesting birds, and the remains of caribou and seals killed by wolves and polar bears. They are very well adapted to survive harsh climatic conditions. Their long, thick, dense fur provides remarkable insulation. Body heat also is retained because limbs are short and ears are small to reduce the amount of surface area exposed to the weather, and the footpads are furred. Unlike other canids, fur color varies seasonally. It is white in the winter and a grayish brown in summer. Arctic foxes are trapped for their fur and also raised on commercial fur farms. They were introduced onto many of the Aleutian Islands in the late 1800s as a fur resource, but because they devastated coastal bird populations, efforts since then have focused on eliminating arctic foxes from these islands.

There is some question whether kit and swift foxes are really different species, but genetic evidence suggests that they are. Their ranges do not overlap, with kit foxes found in arid and shrubby areas throughout much of the southwestern United States and southeastern Oregon to west Texas and Mexico. Swift foxes occur throughout the central plains states from southern Canada to north Texas. Both species are about the same size and are very similar in appearance. They are smaller than red or gray foxes, and have grayish-tan pelage above and lighter undersides. Kit foxes have longer ears, narrower muzzles, and relatively longer tails than swift foxes. Both species are nocturnal, hunting rodents, rabbits, insects, birds, and reptiles, and spend their days in dens protected from heat. Kit and swift foxes are much reduced in density and distribution because of trapping, loss of habitat, and predator control programs aimed at coyotes. Neither species has ever been significant in the fur market, although they are trapped in several states. In other states, however, both species are listed as threatened or endangered.

Island foxes are closely related to the gray fox, but are a separate species because of conditions provided by twelve thousand years of isolation on six of the Channel Islands, the only place they occur. These are very small foxes, only about half the size of gray foxes but with the same pelage coloration. They are found in various habitats and are active both day and night, feeding on mice, birds, insects, and plant material. Because of their limited distribution, they have never been a factor in the fur market. Island foxes have been removed or relocated from San Clemente Island because they prey on an endangered subspecies of bird, the San Clemente loggerhead shrike. The island fox also was recently added to the federal Endangered Species List.

CATS

The order Carnivora comprises nine families in North America, with one of the most familiar being the cats (Felidae). There are seven species of felids in North America, only three of which commonly occur (although they are rarely seen) north of Mexico: the mountain lion, lynx, and bobcat. The largest North American cat, the jaguar, is rare in Mexico and Central America and essentially extirpated from the United States; only occasionally does a vagrant cross the border. The endangered ocelot is a rare species in southern Arizona and Texas. Ocelots may occasionally enter Big Bend National Park, but they have never been confirmed there. Likewise, the jaguarundi, also rare in southern Arizona and Texas, has been reported from Big Bend but never confirmed. A fourth species of cat is the endangered, little-known margay. This smallest of North American felids has rarely, if ever, occurred as far north as south Texas.

All members of the cat family have short, rounded muzzles. Ears are generally small and rounded. Unlike other carnivores, the sharp, curved claws are retractile. All cats—including the family pet—hunt by stalking their prey and pouncing after a short chase. Teeth and jaws are strong and built for taking live vertebrate prey. Cats have stout bodies and relatively long legs and are excellent climbers.

Mountain Lion

Seeing a wild mountain lion, one of North America's majestic large predators and a symbol of wilderness, provides a lasting memory for the few fortunate people who experience it. Historically, no species of mammal in the New World had a more extensive distribution than did the mountain lion. It ranged from northern British Columbia coast to coast across Canada and the United States and south to southern Chile and Argentina. Like the wolf and the grizzly bear, the mountain lion was extirpated throughout much of its natural range concurrent with European settlement. Today, there probably are no longer viable populations in the eastern United States except in southern Florida.

Mountain lions are large cats with long dark-tipped tails. The short pelage is usually yellow brown, sometimes gray or reddish brown, with a paler belly. (The species name *concolor* is Latin for "together color" denoting the generally uniform

Mountain lions are solitary hunters that prey on ungulates such as mule deer and elk. Greatly reduced throughout much of their historic range, mountain lions occur on several of the western national parks. A small population also occurs in Everglades National Park, where they are locally known as Florida panthers.

coloration.) Mountain lions vary in size and weight but are larger than bobcats and lynx and have a much longer tail. Males are larger and can be 50 percent heavier than females. Communication involves a variety of facial expressions and several vocalizations, and mountain lions are the largest cats that purr.

Solitary hunters, mountain lions wander throughout their extensive home ranges. They are the largest predators throughout most of their current range. They are active primarily at dawn and dusk, although they may hunt any time of the day or night. Mountain lions do not make permanent den sites but spend a few days under downed logs, in brush thickets, or in abandoned dens before moving. Various habitats are used for hunting, denning, and resting, but generally mountain lions prefer areas with rugged, broken terrain and heavy brush cover. Cover is necessary for hiding and climbing. Like domestic house cats, mountain lions hunt by stealth and stalking, where cover is essential, with a final dash and pounce. Mule deer are the primary prey of mountain lions. Other ungulates are taken including white-tailed deer, elk, and mountain sheep. Rabbits, smaller carnivores, and rodents are preyed upon opportunistically, but are a small percentage of the diet.

After killing an ungulate, by bites to the neck and throat, a mountain lion will gorge itself, then cache the remainder of the carcass, covering it with grass, sticks, and other debris. It will den nearby and continue to feed off the carcass for several days. The mountain lion does not feed on carrion, however. A mountain lion takes about one deer a week, although a female with kittens may need one every few days. Regardless, mountain lion predation on deer and elk has no impact on healthy populations, but actually may serve a positive function to dampen extreme fluctuations in numbers. Livestock such as calves and sheep are also taken, especially when deer densities are low.

Breeding can occur throughout the year. Pairs remain together for only about a week, and females raise their kittens alone. Following gestation of 3.5 months, an average of two or three kittens is born. Newborns weigh about one pound and have a yellow-brown coat with dark spots, and a ringed tail. They begin eating solid food by six weeks of age, but continue to nurse until they are two to three months old. Young stay with their mother until they are twelve to eighteen months old, so an adult female usually breeds only every other year. When they disperse, juvenile males usually travel much farther than do young females. Mountain lions have few natural predators and may live ten to fifteen years in the wild.

CONSERVATION CONCERNS

Within the contiguous United States, mountain lions inhabit only about 30 percent of their historical range. Loss of habitat, overharvest, and targeted control programs effectively eliminated them. There are no known populations of mountain lions in the east except for thirty to fifty endangered Florida panthers (a subspecies of mountain lion) in Everglades National Park, Big Cypress National Preserve, and the surrounding area. Loss of habitat and mortality from cars make survival of this population tenuous at best. Increased sightings of mountain lions in Minnesota, and in western North Dakota, Nebraska, and Oklahoma, suggest potential reestablishment may be expanding eastward. Increased distribution and density of mountain lions may be related to the increasing number of white-tailed deer throughout most of the eastern United States. However, because mountain lions are seen so infrequently, even where they are common, it is difficult to confirm their presence visually. Evidence of occurrence is often based on tracks or other sign.

Mountain lions are legally hunted in many of the western states and most mortality is due to hunting. Several states have made it more difficult to take lions by restricting the use of dogs to tree them; California recently banned mountain lion

Mountain Lion *Puma concolor*	
COMMON NAMES	Cougar, Panther, Puma
HEAD AND BODY LENGTH	54 inches (138 cm) males
	48 inches (123 cm) females
TAIL LENGTH	31 inches (79 cm) males
	28 inches (71 cm) females
BODY WEIGHT	156 pounds (71 kg) males
	97 pounds (44 kg) females

hunting. Interestingly, since the reintroduction of wolves to Yellowstone National Park, competition may have forced mountain lions to change some of their movement and hunting patterns. Populations of mountain lions in the western United States are stable and, as noted, a slow expansion eastward may be occurring. As with other large predators that are potentially dangerous to people, the future conservation and management challenge is to attain a balance between viable populations of mountain lions and ever-increasing human encroachment on their diminishing wilderness habitats.

Lynx and Bobcat

Bobcats and lynx are both medium-sized, short-tailed spotted cats. There is a long history of uncertainty among scientists over the genus name for these species, with both *Felis* (Latin for cat) and *Lynx* (Greek for cat) used in the literature. The lynx is distributed throughout Alaska and most of Canada, and in northern tier states in the Northeast, the Midwest, and the Rocky Mountains south to Colorado. The bobcat occurs from southern Canada throughout most of the contiguous United States, but with populations very restricted in central midwestern states from Iowa to Ohio. There is little overlap in the geographic ranges of lynx and bobcats. Both species are rarely observed, but park visitors are more likely to glimpse a bobcat along a roadside; unlike lynx, they are more often near developed areas.

Bobcats and lynx have compact, muscular bodies, relatively long legs, and short ("bobbed") tails. They also have ear tufts and long facial hairs that form a flared, downward-pointed beard or "ruff" on each side of the head. The long thick fur of the lynx is a grayish brown, with a lighter chin and belly. Spotting, if evident, is much less distinct than in the bobcat. The bobcat has a more reddish coat color (the species name means red) with much more discrete dark spots or streaks. Compared to the bobcat, the lynx has longer black ear tufts and a longer, more conspicuous facial ruff, and the tip of the tail is completely circled in black, not just the top as in the bobcat. The lynx also is slightly larger than the bobcat, with longer legs and disproportionately large, round heavily furred feet that act as snowshoes. Both species molt just once annually.

Lynx are closely associated with northern coniferous forests, often at high elevation, with deep winter snow. Preferred habitats have a mix of old growth and younger stands of trees, thick under-story vegetation, abundant downed woody debris, rugged terrain, and rocky outcrops. Bobcats are much more habitat generalists. They use almost any habitat that has abundant underbrush and rugged terrain that provides cover for stalking prey. Dens of lynx and bobcats are under uprooted trees, in hollow logs, in rock fissures or ledges, or in small caves, and are not permanent.

The dependence of lynx on snowshoe hares as primary prey is well known, and they make up 80 percent or more of the diet. The large feet of lynx allow them to move across snow effectively to take the hares. Just as snowshoe hares exhibit well-defined ten-year population cycles (though no one knows why) so do lynx, with a one- or two-year lag that follows increases or decreases in hares. When snowshoe hare populations are low, lynx rely on ptarmigan, ruffed grouse, red squirrels, muskrats, and mice to survive. When hare populations are high, excess food may be cached under snow or woody debris. Home range size of lynx also varies depending on hare population density, with larger areas necessary to find sufficient food when hare populations are low. Home range size of bobcats varies as well, depending on prey density, sex, and season. Cottontail rabbits and hares make up a large percentage of a bobcat's diet, with small mammals, ground-nesting birds, and deer fawns also taken. Populations of bobcats do not fluctuate like those of lynx, however. Neither bobcats nor lynx chase down prey, but hunt by stealth and ambush, waiting to surprise and pounce on unwary prey. Bobcats and lynx are active at dawn and dusk (crepuscular) and throughout the night. Other than the breeding season, and adult females with kittens, individuals are solitary. Scent and scrapes are used to mark territorial boundaries.

Reproduction in lynx and bobcats is similar. They generally breed in late winter or early spring, depending on geography. Males are with females only to mate, and pairs do not stay together. Females have a single litter per year. Gestation is nine to ten weeks, and average litter size is three or four kittens. Although kittens are fully furred with spotted pelage, their eyes are closed and they are completely dependent. They begin eating meat when about one month old. Lynx kittens continue to nurse for another five months, and stay with their mother until

Distinctly spotted reddish fur, reduced facial "ruff," and short ear tufts all serve to distinguish this bobcat from the closely related lynx. Bobcats and lynx are highly secretive species rarely seen by park visitors.

the following breeding season, learning how to hunt. Bobcat kittens are weaned at two months of age and disperse in the autumn. When snowshoe hare populations are low, fewer adult female lynx breed. If they do reproduce, a large percentage of kittens starve. Wolves, mountain lions, or wolverines may prey on lynx and bobcats. Most mortality, however, is through starvation or trapping for furs.

CONSERVATION CONCERNS

Lynx have been trapped throughout Canada for more than two hundred years, and the records of pelts purchased by the Hudson Bay Company show their cyclic population fluctuations. Today, harvest for furs continues, but with strict limits in some areas because of concern over declining populations. Another serious concern is loss of wilderness habitat. Human access by snowmobiles and the proliferation of roads for development of energy resources can negatively affect lynx.

Thirty years ago, almost no research was done on either lynx or bobcats, and they were of minor concern to wildlife managers. Since the 1975 Convention on International Trade in Endangered Species (CITES), all states have greatly expanded research programs on bobcats to show that trapping and subsequent export of furs was "no detriment" to populations. Bobcats are now trapped as a furbearer in thirty-eight states, and populations are considered stable or increasing. Nine other states protect bobcats from any trapping. Historically, every state paid a bounty to help eradicate bobcats, as they did with wolves, bears, mountain lions, coyotes, and other predators. The ineffectiveness of bounties as part of a predator control program is now clearly recognized, although several states continued bounties on bobcats until recently. Declines in population necessitated reintroduction of bobcats in many states, and these generally have been successful.

BOBCAT	
Lynx rufus	
COMMON NAMES	Wildcat, Bob-tailed cat, Lynx cat, Red lynx
HEAD AND BODY LENGTH	26–30 inches (66–76 cm)
TAIL LENGTH	4–5 inches (10–13 cm)
BODY WEIGHT	20–30 pounds (9–14 kg)
LYNX	
Lynx canadensis	
COMMON NAMES	Canada lynx, Gray wildcat
HEAD AND BODY LENGTH	32–36 inches (81–91 cm)
TAIL LENGTH	4–5 inches (10–13 cm)
BODY WEIGHT	20–35 pounds (9–16 kg)

BEARS

There are nine species of bears (family Ursidae) worldwide; three occur in North America: the polar bear, grizzly bear, and black bear. Polar bears have a circumpolar distribution on Arctic ice and coastlines in the Eastern and Western Hemispheres. They are not far enough inland in Alaska to occur in any of the national parks, however. Black bears and the larger grizzly bears are the only ursids found in national parks, and both species occur in several parks. Black and grizzly bears share many life history characteristics in terms of feeding, dens, activity, dormancy, and reproduction.

Black Bear

Black bears are common in several national parks but rare and infrequent visitors in others. Nonetheless, park visitors are much more likely to see black bears than the less common grizzly bear. Several physical features serve to distinguish black bears from grizzlies. The black bear is smaller, without a noticeable hump between the shoulders like the grizzly. The ears are longer and the profile of the face and nose is straight or slightly convex, not concave or dish-shaped as in the grizzly. The nonretractable claws of the black bear are only about one-fourth as long as those of the grizzly bear and are not as strongly curved. The foot pads, and resulting tracks made by each species, are also different. The long, dense, glossy fur of many black bears is actually black, especially in eastern areas, and is not light-tipped or "grizzled." Fur color is highly variable, however, and pelage of western bears may be brown, cinnamon, gray, or cream. The stocky, powerful build and flat-footed gait of black bears are familiar to everyone. Built for strength, they nonetheless can run thirty miles per hour for short distances and are good swimmers. Average weight of males (boars) is about twice that of females (sows).

Black bears occur from Alaska through Canada and many western and northern tier states. They have a patchy distribution in many other regions of the United States but are absent from much of the Midwest from North Dakota to Texas. Usually associated with deciduous or coniferous forests, wet meadows, and dense brush cover, black bears are exceptionally adaptable and occur in a variety of habitats. As long as food and cover are available, populations can thrive from northern tundra to suburban housing developments.

Black bears are solitary, except for females with young. Home range size of males is much too large to actively defend, and individuals just try to avoid each other. Individual bears mark trees with scent, rubbing, and claw marks, probably to facilitate mutual avoidance. These "bear trees" or "register trees" are obvious indicators of bears in an area. Although classified as a carnivore, black bears are quintessential omnivores and will eat almost anything. Most of their diet is vegetation, berries, nuts (especially acorns), and roots. Their fondness for bees and honey is so well known it is a theme in children's literature. Black bears also take meat, often in the form of carrion, as well as insects and small vertebrates. They are frequent visitors to garbage dumps and camps. Black bears have poor eyesight, but their hearing and sense of smell are excellent.

Black bears are active throughout the night and early morning. Shallow depressions in the ground or under a stump serve as daytime resting sites. Permanent dens are under fallen trees, in shallow caves, or dug into hillsides. Black bears are not true hibernators. After putting on extensive body fat during summer and fall, they pass harsh winters by becoming inactive. Physiologically, their body temperature, respiration, and heart rate all decrease but not as much as in true hibernators. This period of dormancy is often called "winter lethargy" or "winter sleep," and bears are able to arouse more quickly than true hibernators. Nonetheless, they do not eat, drink, or eliminate wastes but live for many months entirely off stored body fat. In the far north, both sexes become dormant during winter; in the south only females do.

Reproduction is also tied to winter dormancy. Mating occurs in early summer, but implantation and actual development are delayed until November. True gestation is only six to eight weeks, which is very short for a large mammal. Birth occurs in January while the female is dormant in the den. Average litter size is two to three highly altricial cubs. Newborns weigh less than a pound. Other than marsupials, no mammal has a larger ratio of mother-to-young body weight (about 250:1). By the time cubs emerge from the den in the spring, they weigh twelve to fifteen pounds. Weaned at seven months of age, cubs and mother stay together throughout the next winter. Young disperse when they are seventeen to eighteen months old. Both sexes probably do not successfully breed until they are five or six years old; females then only breed every two or three years.

CONSERVATION CONCERNS

Bears can be a problem when drawn to garbage or improperly stored food at campsites. All parks stress a number of precautions regarding food storage, garbage disposal, and other activities of campers and hikers, to minimize and reduce negative interactions between people and bears. Everyone should know that human encounters with black bears are dangerous, especially when highly protective females are with young cubs. Black bears are generally shy and reclusive, however; they usually retreat from people and unprovoked attacks are very rare. Data from the National Center for Health Statistics help reinforce this fact. Considering the number of people killed by black bears in the past hundred years, sixty times more individuals have been killed by domestic dogs, 180 times more by bees, and 350 times more people have died from lightning strikes. Black bears are a popular game animal, with up to fifty thousand legally harvested annually throughout North America during fall hunting seasons. Unfortunately, close to this number may be poached each year for their meat, hides, gallbladders, and paws. Illegal killing can have significant negative impacts on smaller, remnant populations of black bears. Although populations in most areas appear to be stable, the low reproductive potential of black bears, resulting from small litter size, long time to sexual maturity, and long interval between litters, makes them especially vulnerable to overharvest, illegal killing, and habitat loss.

Black bears are adept tree climbers. Smaller than grizzlies, black bears are much less carnivorous but will eat meat when necessary.

BLACK BEAR *Ursus americanus*	
COMMON NAMES	American black bear, Cinnamon bear, Glacier bear, Kermode's bear
HEAD AND BODY LENGTH	5.6 feet (1.7 m) males 4.6 feet (1.4 m) females
TAIL LENGTH	4 inches (10 cm)
BODY WEIGHT	264 pounds (120 kg) males 176 pounds (80 kg) females

Grizzly Bear

Few species symbolize size, strength, and power as much as the grizzly bear. Like other large carnivores, grizzlies were extirpated from most of their historic range in the contiguous western United States and Mexico concurrent with European colonization. Today, grizzlies remain throughout Alaska and in northwestern Canada to Hudson Bay, with remnant populations in northwestern Washington, Montana, and Wyoming.

Biologists in the early 1900s had a difficult time determining the number of species and subspecies of grizzly bears. Resulting taxonomic exuberance produced over a hundred named taxa. Some confusion remains today regarding common names. Grizzly, brown, and Kodiak bears are all the same species, although there are noticeable size differences associated with location, habitat, and diet of different populations. Biologists now recognize five distinct genetic groups (clades) of grizzly bears throughout the world, but again, only one species.

Body weights of grizzlies depend on region, season, and age. Males are about 10 percent bigger than females. Where they occur in the same areas, grizzly bears can be confused with black bears. Larger in size and weight than black bears, grizzlies have a huge head with a concave facial profile, a large hump between the shoulders, and light-tipped or frosted ("grizzled") fur, especially along the back and shoulders. Black bears have none of these features. Grizzly bears also have much longer, more noticeable claws on the forefeet than black bears do. Pelage color usually is brown but can vary from a yellowish tan to nearly black, with a lighter muzzle.

Grizzlies once occurred in all habitats of western North America, but because of human pressure they are now restricted to more remote tundra, alpine meadows, woodlands, and grasslands. Home ranges are exceptionally large. Males may use areas of 150 square miles or more, which overlap the ranges of several females. Grizzly bears are somewhat more social than black bears. When several grizzlies are in the same area, aggressive interactions are reduced because of the dominance hierarchy: adult males are dominant, then adult females with cubs, while young animals are submissive.

Grizzly bears exhibit many of the same life history characteristics as black bears. Grizzlies are opportunistic omnivores and will eat practically anything organic. Feeding habits shift seasonally, but vegetation makes up a large part of the diet, including a wide assortment of grasses, fungi, fruits, nuts, roots, and berries. Grizzlies are powerful enough to prey on any terrestrial mammal in North America, including bison. Ungulates taken are usually the sick or old, or they are eaten as carrion. Small mammals, including ground squirrels and marmots, are dug from burrows, and insects, larvae, and other invertebrates are also consumed. Spawning runs of fish will draw several bears to a shallow spot in the river. Like black bears, grizzlies are active at night and early morning, rely on keen senses of smell and hearing, and are good swimmers. Unlike black bears, however, they do not climb trees.

Winter dens are excavated and oriented so they will be free of water from melting snow. Grizzly bears have the same cycles of winter dormancy and reproduction as black bears, and the same need to put on a great deal of fat each summer and fall. Grizzlies may put on up to four hundred pounds of fat before entering a den, where they remain dormant for up to seven months. Although they are not true hibernators, during "winter sleep" or "winter lethargy" they decrease body temperature and heart rate.

Breeding takes place in early summer. Females are promiscuous and mate with several males. The fertilized eggs do not implant in the uterus until November, when the dormant female dens for the winter. Actual gestation is only six to eight weeks. Two or three helpless tiny cubs, about ten inches long and one pound, are born in the den. Cubs remain with their mother during a prolonged lactation of 1.5 to 2.5 years, and are forced to disperse once she breeds again. Females have their first litter when six or seven years old. Reproductive intervals vary from 2.5 to 4.5 years thereafter.

CONSERVATION CONCERNS

As noted, grizzly bears are much reduced throughout their historic range. Current distribution in the contiguous United States is about 1 percent of their former range. Given their size, the amount of area they need for foraging, and human population densities, there are few remaining areas that can accommodate grizzly bears. Also, considering their low reproductive rates, grizzly bear populations are

These grizzly bears in Katmai National Park wait patiently for their next meal of spawning fish to swim upstream. Visitors spend millions of dollars annually in Alaska to view and photograph grizzly bears in the wild.

GRIZZLY BEAR *Ursus arctos*	
COMMON NAMES	Brown bear, Kodiak bear, Range bear, Silver tip
HEAD AND BODY LENGTH	6–7 feet (1.8–2.1 m)
TAIL LENGTH	4 inches (10 cm)
BODY WEIGHT	330–760 pounds (150–345 kg) males 220–495 pounds (100–225 kg) females

slow to recover from overharvest or habitat loss. Hunting is allowed in Canada and Alaska. An average of sixteen hundred grizzly bears is taken annually, most of them in Alaska. Throughout their range, there is additional human-caused mortality of grizzlies from defense-of-life incidents. Bear hunting generates a great deal of income but not as much as recreational viewing and photographing grizzly bears, which amounts to tens of millions of dollars annually. South of Canada, hunting grizzly bears is illegal. They are currently listed as threatened under the Endangered Species Act.

Outside of Alaska, grizzly bears now occur in three national parks. Secure populations exist only in Alaska, however. One of the primary causes of declining populations is habitat fragmentation from roads. Behaviorally, grizzlies are known to avoid roads. In addition, roads and increased access are directly related to the number of grizzly bears harvested by hunters. State and federal management agencies, private citizens' groups, and interagency planning teams have formulated recovery plans to increase population densities of grizzly bears in the United States and Canada. Recovery zones are the Northern Continental Divide Ecosystem, which includes Glacier National Park, and the Greater Yellowstone Ecosystem, which includes Yellowstone National Park. Within each zone, and inclusive Bear Management Units, target densities have been established for grizzly bears based on the average number of females with cubs. A plan was also worked out among state agencies, the U.S. Forest Service, the U.S. Fish and Wildlife Service, and citizens' groups to reintroduce grizzly bears to the wilderness areas of central Idaho and eastern Montana. The Department of Interior under the George W. Bush administration, however, refused to fund the initiative. Grizzly bears also are one of the focal species in a proposed conservation plan called the Yellowstone-to-Yukon Biodiversity Strategy. This plan was conceived to protect existing areas, such as the national parks and other public lands, and to restore ecosystem health with interconnected habitat corridors from Yellowstone National Park to the Yukon.

As with black bears, precautions to minimize conflicts between people and grizzlies are well known and publicized, including not feeding bears. Campers and backcountry hikers should stay in a group, wearing "bear bells" to alert the animals to human presence, pack food in airtight containers, and not leave garbage accessible. Restrictions on hiking and camping are sometimes in effect in bear country, and other precautions are posted at each park to reduce danger to people and adverse consequences to bears.

RACCOONS

Few people would fail to recognize a raccoon, but they may be much less familiar with two other members of the raccoon family (Procyonidae) that occur in North America: the ringtail and the white-nosed coati. The best-known procyonid, of course, is the raccoon. The stocky body, slightly waddling gait, familiar black face mask and pointed nose, and bushy, black-tipped tail of alternating light and dark bands make the raccoon difficult to confuse with any other species. Pelage is long, with coarse, shaggy, yellowish gray to blackish guard hairs that overlay short, dense underfur. Fur is somewhat darker on the back than on the belly. Males are generally larger and heavier than females. Raccoons are found from central Canada south through central Mexico. They occur throughout the contiguous United States, even moving into higher elevation mountainous areas and arid regions of the southwest. They also inhabit several islands of southeastern Alaska.

Habitat generalists, raccoons are at home in woods, wetlands, marshes, farmlands, brushy old-fields, and residential areas—almost anyplace near a stream, lake, pond, or other source of water. Raccoons are primarily nocturnal, and their omnivorous feeding habits are just as general and opportunistic as their habitats. Depending on season and availability, they consume vegetation, seeds, fruits, insects, carrion, birds, crustaceans, mollusks, small mammals, fishes, and other vertebrates. They take almost anything organic, and are well-known raiders of residential garbage cans. Their forefeet have

RACCOON	
Procyon lotor	
COMMON NAMES	Coon, Northern raccoon
HEAD AND BODY LENGTH	18–26 inches (46–67 cm)
TAIL LENGTH	8–14 inches (20–36 cm)
BODY WEIGHT	13–33 pounds (6–15 kg)

RINGTAIL	
Bassariscus astutus	
COMMON NAMES	Cacomistle, Ringtail cat, Miner's cat, Civet cat
HEAD AND BODY LENGTH	12–15 inches (30–38 cm)
TAIL LENGTH	12–17 inches (30–43 cm)
BODY WEIGHT	2–3 pounds (0.9–1.4 kg)

WHITE-NOSED COATI	
Nasua narica	
COMMON NAMES	Coatimundi, Chulo bear
HEAD AND BODY LENGTH	16–27 inches (41–69 cm)
TAIL LENGTH	14–27 inches (35–69 cm)
BODY WEIGHT	9–13 pounds (4–6 kg)

five digits with short, nonretractable claws, and are highly flexible and sensitive. Raccoons have exceptional manual dexterity. Folklore is often fallible, but the common perception of raccoons "washing" food is often true, provided water is readily available. The species name is Latin for "a washer," and the common name is derived from the Algonquin Indian word for "scratches with his hands." Raccoons are not being hygienic by washing food, but are probably gaining better sensory information about food items by putting them in water.

Raccoons are excellent swimmers and adept climbers, seeking food or shelter in trees. Like the related ringtails and coatis, they are one of very few mammals that are able to descend from trees head-first. Favored den sites are in hollow trees, although rocky areas, cavities under downed trees, and abandoned burrows of marmots or foxes are readily used. Raccoons do not hibernate, but will remain in the den through prolonged periods of inclement winter weather, living off stored body fat. An individual will have several different den sites and periodically move among them. Raccoons are generally solitary, although "piles" of up to twenty-five may share a large communal den, often to conserve body heat during winter. Primary predators include coyotes and bobcats, although predation has a relatively negligible impact on populations compared to disease and human-caused mortality.

Most mating commences in February or March, and males breed with several females. An average of four young (called cubs, pups, or kits) is dropped following a gestation of nine weeks. Newborn raccoons weigh about three ounces but have enough fur that the mask and ringed tail are noticeable. Cubs are weaned at four months of age, when they begin nightly foraging with their mother, but don't reach adult body weight until they are about two years old. They often don't disperse far, but may remain in the natal area to den near their mother and siblings.

Ringtail

Ringtails are distributed throughout the southwestern United States, from Texas west to California and southwestern Oregon. Most people are unfamiliar with them and rarely see this secretive, nocturnal species. In many areas, ringtails may be more common than expected, but often their status is poorly documented. Somewhat slender and graceful, ringtails are about the size of a house cat. The yellowish tan fur is somewhat darker on the back

and shoulders and lighter on the belly. They have large brown eyes framed in white fur. The common name derives from the long bushy tail with fourteen to sixteen alternating light and dark rings. Ringtails inhabit woodlands, arid canyons, and rimrock areas, where they are agile climbers. They are much more territorial than raccoons and scent mark to delimit their areas. They also are much more carnivorous. Ringtails prey on mice, rabbits, and woodrats, as well as other small vertebrates and invertebrates, but also take some fruits, nuts, and berries. Den sites used and reproductive characteristics are very similar to those of raccoons.

White-nosed Coati

White-nosed coatis extend throughout Central America and Mexico, reaching their northernmost distribution in extreme south Texas, southeast Arizona, and southwest New Mexico. Smaller than the raccoon but larger than the ringtail, the coati has dark-brown to black pelage, white eye rings, and a long, flexible snout. The tail is about the length of the head and body, is held erect, and may or may not have faint rings. Coatis are never far from water, often in wooded canyons. Unlike raccoons and ringtails, the white-nosed coati is diurnal, doing most of its foraging on the ground with the tail held in the air. It feeds on small vertebrates, invertebrates, nuts, fruit, and carrion. It

With its chunky body and black mask, the raccoon is easily recognizable. Raccoons are an important part of the fur industry, but unfortunately they are also the primary carrier of rabies in the eastern United States. The droppings of raccoons serve an important ecological function by dispersing seeds.

usually spends the night sleeping in a tree or on a rocky ledge. The coati is the most social procyonid. Size of a "band" varies but may reach forty individuals. Bands are composed of adult females and subadults and juveniles of both sexes. Adult males are generally solitary. Females all mate during about two weeks in April. Gestation is ten to eleven weeks, and females give birth in a tree nest or den. Newborns weigh about six ounces and nurse until about four months of age.

CONSERVATION CONCERNS

Raccoons are taken by hunters and trappers and are one of the most important furbearers in North America. Only muskrats may be taken in greater numbers each year, but raccoons generate the greatest amount of revenue. Throughout most of the 1900s, one to two million raccoons were harvested in the United States annually, depending on the price of furs. The record harvest occurred in the 1979–80 season, when 5.1 million raccoons were taken. Heavy trapping and hunting pressure does not seem to negatively affect raccoon population densities, however. Because of their importance as furbearers, raccoons have been introduced to several countries in Europe and Asia. The positive economic impact of raccoons as a valued fur resource remains today in the United States, but negative economic aspects also loom large. Raccoons are often the primary nuisance species in suburban residential areas. Homeowners nationally spend millions of dollars each year to have raccoons removed from attics and garages. Raccoons have an even greater negative economic impact nationally than as a pest. They are the primary focus for rabies outbreaks in the eastern United States. Tens of millions of dollars are spent annually to monitor, study, and contain outbreaks of raccoon rabies, which are slowly moving westward. A significant factor is the spread of the disease to household pets, especially cats, which can then infect humans. Raccoons enjoy widespread popular public appeal, but many well-meaning people in residential areas feed them without realizing the potential serious health risks.

Ringtails are occasionally trapped and managed as a furbearer in Arizona, New Mexico, and Texas, but are of little economic significance. Population densities are believed to be stable in most areas, but little information exists. White-nosed coatis are a protected species in Texas and New Mexico but are harvested in Arizona. Little is known of the biology of populations in the United States, but densities in northern Mexico have declined significantly.

MUSTELIDS

Of the nine families in the order Carnivora that occur in North America, the weasels (family Mustelidae) are certainly the most diverse. Despite their morphological and behavioral diversity, all mustelids have elongated bodies, relatively short legs, and well-developed anal scent glands. Most are important furbearers. Mustelids include three species of small carnivores commonly called weasels (the long-tailed, short-tailed, and least weasels), the semi-aquatic river otter, the marine sea otter, the mink, the badger, the wolverine, and two species that are probably least familiar to many people—the closely related marten and fisher.

American Marten and Fisher

About the size of a house cat, the American marten (there is a different species of marten in Europe and Asia) is typically weasel-like with a long slender body, fairly short legs, and bushy tail. Its muzzle is more elongated than in the smaller weasels, and it has prominent rounded ears. The marten has thick, dense, silky pelage that is golden brown above and darker below, with irregular cream-colored to reddish splashes of fur on the throat and chest. The fisher occurs only in North America and is much larger, stockier, and heavier than the marten. It is about the size of a fox, with darker brown to black glossy pelage. Ears are large and rounded and the tail is fairly bushy. The head and shoulders have a silver, grizzled wash that is lighter than the rest of the pelage, and the fisher generally lacks the throat patches found in martens.

Martens and fishers inhabit cool, wet, northern coniferous (boreal) forests of fir, spruce, hemlock, and pine. Martens currently range throughout interior Alaska and forested portions of Canada. In the United States, they are in the Northeast, northern Midwest, and higher elevations of Washington, Oregon, California, and the Rocky Mountains. Fishers occur coast to coast across south-central Canada and, like martens, extend into parts of the northeastern, midwestern, and western United States. Both species occur in mature, late successional, or mixed-age forests with closed canopy. A dense understory of downed woody debris, shrubs, hollow logs, and rocky outcrops is essential for den

sites, for protection from deep snow, and as habitat for prey species. Martens and fishers are active throughout the year, and neither species hibernates. They forage opportunistically day or night. Small mammals, especially mice and voles, are the primary food of martens, along with squirrels, rabbits, birds, fruits, and seeds. Fishers have a similar diet, but like mountain lions are particularly adept at taking porcupines. The common name fisher is a misnomer (they take very few fish) and may derive from a Dutch term for "polecat." Other than during the breeding season, martens and fishers are solitary. Size of territories depends on habitat and prey availability, but males have larger areas that may encompass several smaller territories of females.

As in most other mustelids, martens and fishers exhibit delayed implantation; following mating, the fertilized egg does not implant for a prolonged period. In the case of the marten, breeding is in July or August but implantation does not occur until late February. Following a true gestation of about a month, young are born in late March. Fishers are somewhat different. Females become receptive seven to ten days after dropping a litter; that is, they exhibit a postpartum estrus and mate in late March or early April. Implantation is delayed for more than ten months until mid-February. Gestation is one month and, like martens, the young are dropped in late March. Average litter size in both species is about three and young (kits) are altricial. Marten kits are weaned at about six weeks of age and disperse in August. Fishers are weaned when seven to eight weeks old and disperse a month or two later.

CONSERVATION CONCERNS

The dense, lustrous pelage of martens and fishers made them such valuable furbearers that overharvest throughout the 1800s and early 1900s led to drastic declines in density and distribution of both species. Overharvest was exacerbated by loss of habitat through excessive logging, and populations were extirpated in many areas south of Canada. Many of these populations have rebounded due to harvest restrictions, reforestation, and reintroduction programs for martens and fishers in many states where they once occurred. Both species are legally harvested in Alaska and Canada. Martens and fishers are legally trapped in some of the contiguous forty-eight states, but are protected species in others.

Weasels

The slender elongated body shape of a weasel may be familiar to many people, but park visitors will rarely catch a glimpse of one of these small secretive carnivores. Three species of weasels occur in North America and the national parks, and all share similar life history patterns. Long-tailed weasels are the most widely distributed members of the highly diverse weasel family (Mustelidae) in the Western Hemisphere. They are the largest of the three species that are actually called weasels, and occur in southwestern Canada, the United States except for small portions of the arid southwest, and south throughout Mexico and Central America. Short-tailed weasels occur in northern areas around the world, including all of Alaska and Canada, much of the western United States, and the Great Lakes area and the northeastern region. Least weasels also occur worldwide in northern areas including Alaska, Canada, and much of the north-central and eastern United States.

The marten superficially resembles a house cat. It is a fairly common inhabitant of northern coniferous forests, where it is active throughout the year despite cold weather and heavy snowfall.

AMERICAN MARTEN	
Martes americana	
COMMON NAMES	Pine martin, Martin, American sable
HEAD AND BODY LENGTH	14–18 inches (36–46 cm) males
	12–16 inches (30–40 cm) females
TAIL LENGTH	8–9 inches (20–23 cm) males
	7–8 inches (18–20 cm) females
BODY WEIGHT	1.1–2.6 pounds (0.5–1.2 kg) males
	0.7–1.8 pounds (0.3–0.8 kg) females

FISHER	
Martes pennanti	
COMMON NAMES	Fisher cat, American sable
HEAD AND BODY LENGTH	21–31 inches (53–79 cm) males
	17–23 inches (44–59 cm) females
TAIL LENGTH	15–16 inches (38–41 cm) males
	12–14 inches (31–36 cm) females
BODY WEIGHT	8–12 pounds (3.6–5.5 kg) males
	4.4–5.5 pounds (2–2.5 kg) females

The white fur of this short-tailed weasel helps camouflage it during a snowy winter in Grand Teton National Park. During summer, fur color is brown. Weasels do not hibernate but remain active in search of prey throughout the year.

With flat heads, small black eyes, long thin bodies, and short legs, weasels are well suited to getting through small tunnels, burrows, gaps, or crevices to catch prey. Males are larger and heavier than females in all three species. Female long-tailed weasels are about the size of male short-tailed weasels. Least weasels are the most diminutive. About the size of a man's thumb, they are the smallest carnivores in the world. All have rich brown fur on the back, with a white chin. Long-tailed weasels have yellowish pelage on the throat and belly during the summer; short-tailed and least weasels have white belly fur. Pelage color changes with seasonal molts. During the winter, fur becomes entirely white, at least in northern areas with snow cover, and provides effective camouflage. Long-tailed weasels are aptly named because tail length nearly equals head and body length. As might be expected, the tail is somewhat shorter in short-tailed weasels, whereas in least weasels it is less than one-quarter of the head and body length. The tail has a black tip throughout the year except in least weasels, where it is entirely brown.

There are few elevations or habitats throughout their vast range in which long-tailed weasels cannot be found. They inhabit alpine areas, forests, grasslands, farmland, marshes, and hedgerows, usually close to water. Short-tailed and least weasels occur in many of the same habitats, as well as tundra. All weasels are active throughout the year and are highly carnivorous. Long-tailed weasels, because they are larger than the other species, can take larger prey such as rabbits, hares, and squirrels, although for all weasels the vast majority of the diet is smaller mice, voles, birds, frogs, snakes, insects, and carrion. Short-tailed weasels also attack prey larger than they are, including rabbits and pikas, but are especially adept at tunneling under snow cover to hunt voles. Weasels have reputations as savage killers, but it is a misconception that they suck the blood from their prey vampire-fashion. They do, however, lap up blood from freshly killed prey. Because of their slender body shape, weasels lose heat quickly. This necessitates a much higher metabolic rate than other species of comparable body weight. As a result, weasels need to consume the equivalent of about 40 percent of their body weight daily in food. Nonetheless, at times weasels may kill more animals than they can consume, and carcasses are then cached in the den for later use.

Dens are in rock piles, hollow logs, or burrows taken over from previously consumed prey. They are lined with leaves, fur, or feathers for insulation against severe winter weather. Like most other mustelids, weasels are solitary until the breeding season. Anal scent glands produce a strong odor that individuals rub over rocks and logs to identify their territory and use for defense. Coyotes, foxes, hawks, owls, snakes, and domestic dogs and cats are among the many potential predators of weasels.

Breeding in long-tailed and short-tailed weasels occurs in late July and August. Implantation of fertilized eggs is delayed for 7.5 months, however, until February or March. Actual gestation is only about twenty-four days, after which four to nine altricial young (pups) are born in the spring. Newborn weasels weigh only about 0.1 ounce, but they grow rapidly and are weaned at about three months of age. They are fully mature by autumn when they disperse from the natal area, after learning how to hunt. Least weasels are quite different. They breed throughout the year and, unlike other mustelids, do not delay implantation. Gestation is about five weeks, and several litters may be produced at any time of the year.

CONSERVATION CONCERNS

Historically, thousands of the white winter pelts of short-tailed weasels, called ermine, graced the robes of European royalty. Today, weasels are of limited economic importance as furbearers, although the

LONG-TAILED WEASEL *Mustela frenata*	
COMMON NAMES	Stoat, Bridled weasel
HEAD AND BODY LENGTH	8–11 inches (20–28 cm)
TAIL LENGTH	3–6 inches (8–15 cm)
BODY WEIGHT	6–15 ounces (170–425 g) males 3–9 ounces (85–255 g) females

SHORT-TAILED WEASEL *Mustela ermina*	
COMMON NAMES	Ermine, Bonaparte's weasel
HEAD AND BODY LENGTH	6–10 inches (15–25 cm)
TAIL LENGTH	2–3 inches (5–7.6 cm)
BODY WEIGHT	3–5 ounces (85–142 g) males 1–3 ounces (28–85 g) females

LEAST WEASEL *Mustela nivalis*	
COMMON NAMES	Dwarf weasel, Pygmy weasel, Mouse weasel
HEAD AND BODY LENGTH	5–7 inches (13–18 cm)
TAIL LENGTH	1–1.6 inch (2.5–4 cm)
BODY WEIGHT	1.4–2.1 ounces (40–60 g) males 1–1.7 ounce (28–48 g) females

white winter pelts are still valuable. Many states have not kept fur sale records for weasels since the 1980s. Unfortunately, when small mammal populations are low, weasels may occasionally raid chicken coops. The benefits of weasels, including all the rodents they take that otherwise would damage crops, certainly outweigh occasional depredations. Often perceived as rare because of their small size and secretive habits, populations of these three species are probably fairly common throughout most of their ranges.

Sea Otter

One of the more charismatic mammals in the national parks, the sea otter is an integral part of coastal marine ecosystems. It is the largest member of the weasel family (Mustelidae), the only one that is almost totally aquatic, but one of the smallest marine mammals. Historically, sea otters occurred in shallow coastal waters of the North Pacific from northern Japan to central Baja California. Hunted for their fur for hundreds of years, they were practically driven to extinction until populations were protected in 1911 and recovery efforts eventually began.

Like most mustelids, male sea otters are larger and heavier than females. Those in northern waters are larger than otters in the south. Forelimbs are short and the sensitive paws, used to forage and manipulate items, have claws. The webbed hind feet are about eight inches long and look like flippers. The tail, about half as long as the head and body, is slightly flattened to aid in swimming and diving. Fur color is light brown to black over most of the body and whitish yellow on the face. Sea otters have the densest pelage known. Long guard hairs overlay underfur with greater than 100,000 hairs per square inch over most of the body. Because sea otters do not have a layer of fat for insulation, they depend on this thick, heavy fur, and the insulating layer of air it traps, for buoyancy and protection against cold water. On the rare occasions when sea otters leave the water and haul out on land or ice, it is often to groom and maintain their fur.

Sea otters inhabit relatively shallow coastal waters, diving as deep as three hundred feet to take clams, crabs, abalone, sea urchins, and other marine invertebrates from the bottom. Slow-moving fishes make up the remainder of their diet. Probably best known for their use of a tool, sea otters place a rock or other hard object on their chest. Prey items are brought to the surface to eat, and as the sea otter floats on its back, it strikes mollusks and crustaceans against the rock to help pry them open. When the sea otter dives again, it holds the rock under an armpit. Because sea otters are small marine mammals and are constantly in cold water, they must maintain a very high metabolic rate to compensate for heat loss—even with the insulation afforded by their dense fur. As a result, they must consume about 30 percent of their body weight daily. Unlike most marine mammals, sea otters do not migrate seasonally. They are more social than most mustelids and large "rafts" of a hundred or more sea otters may form as they rest in the water. Adults segregate, however, as males establish territories to exclude other males.

Births in northern sea otter populations are mainly in late spring. In southern populations, they can occur any month of the year. Females usually first breed when they are three years of age. Males generally are six years old or more before they can establish breeding territories. Mating occurs in the water. Birth is six months later, including a delay in implantation of the fertilized egg and gestation, and a single pup is born. Pups weigh about four pounds and nurse for about six months. Young males disperse, although young females often remain in the natal area. Sea otters are preyed upon by killer whales, white sharks, and grizzly bears, but can live fifteen to twenty years.

CONSERVATION CONCERNS

Their extremely thick fur made sea otters a prized resource for two hundred years, and indigenous people still legally harvest them. Once almost extinct, sea otter populations have rebounded to a large extent in much of their range. In the Western Hemisphere, sea otters from Amchitka and Prince William Sound, Alaska, were successfully reintroduced to coastal areas of southeastern Alaska and Washington. Current threats to populations of sea otters are from coastal development and pollution, oil spills, and entanglement in nets and plastic. Sea otter populations declined significantly following the *Exxon Valdez* oil spill and have yet to fully

With their extremely dense pelage, sea otters remain comfortable in even the coldest water as they float on their backs to crack open shellfish. Sea otters are a key component of healthy coastal ecosystems.

SEA OTTER *Enhydra lutris*	
HEAD AND BODY LENGTH	37 inches (94 cm) males
	32 inches (81 cm) females
TAIL LENGTH	18 inches (46 cm) males
	16 inches (41 cm) females
BODY WEIGHT	40–100 pounds (18–45 kg) males
	24–73 pounds (11–33 kg) females

recover in Prince William Sound. Likewise, the sea otter population in the Aleutian Islands is declining. Conservationists and managers also face ongoing conflicts with commercial and recreational shellfisherman, who compete with sea otters for many of the same resources. Sea otters are an integral part of coastal marine ecology. Through their foraging activities, they reduce population densities of sea urchins, which otherwise would multiply and destroy kelp beds. The food and habitat provided by the growth of dense kelp beds allows a rich diversity of invertebrates, birds, and mammals to remain and flourish in coastal ecosystems.

The river otter is highly adapted to catching fish and other prey in lakes, streams, and marshes from Alaska to Florida. Its thick, durable pelage is highly prized in the fur industry.

Northern River Otter

Best known for their playful antics, chasing and wrestling with each other and sliding down hillsides into the water, river otters are one of the most alluring members of the weasel family (Mustelidae). Historically, river otters were distributed throughout Alaska, Canada, and the contiguous United States wherever there were marine or freshwater habitats and sufficient food. Populations drastically declined and were extirpated from much of the range in the contiguous United States, however, because of overharvest for the fur trade and pollution of rivers, marshes, streams, and estuaries. River otters have since made a comeback in many regions, largely through reintroductions and other conservation efforts.

River otters are one of the largest mustelids, and are highly adapted for their semi-aquatic lifestyle. Small eyes are set high on the slightly flattened head. The elongated body is streamlined, with small ears and no constriction at the neck, so otters move through the water with minimal turbulence. The limbs are short and muscular, feet are webbed, and the long tapered tail is used as a rudder. Pelage is dark brown above, although it may appear black when wet, with a silvery wash on the throat and chest. The fur is very dense and thick, and traps a layer of air that provides extra insulation and buoyancy during dives. River otters are fast and agile swimmers and divers. They move swiftly through the water by undulating motions of the body and tail, can dive forty feet deep, and are quick enough to catch rough fish like carp and suckers, their primary prey. Crayfish, turtles, bird eggs, muskrats, and small mammals round out the diet.

Dens are in old beaver lodges or river banks, under logs, in thickets or rock piles, or in abandoned burrows of coyotes or foxes if they are close to water. A tunnel leads to a nest chamber that is always above water, and is lined with vegetation and fur. Like all mustelids, river otters have anal scent glands and deposit a strong musk to mark occupied territory. Because their short legs and long bodies make movement on land more difficult than in water, otters establish slides in mud, grass, or snow to go down riverbanks or steep slopes. Similar to toboggan runs, otters may slide down head or feet first, on their belly or back—and seem to have fun doing it.

Breeding occurs in early spring. Otters mate in the water and a male may breed with several females. Development of the embryo takes about sixty days, but implantation is delayed for eight to nine months, so young are born in March or April. Females have one litter a year of two to three young. Although newborns (pups) are fully furred at birth, they are about four inches long and helpless. Pups nurse on milk rich in fat and are weaned by about three months of age. Young river otters are often reluctant to enter the water, and both parents must teach them how to swim. River otters are more social than most mustelids. Groups often are composed of a mature female and her young. The young either disperse in the autumn or remain with their mother until spring.

CONSERVATION CONCERNS

Their thick fur is the standard against which other furs are judged for durability, and river otters have been harvested for their pelts for hundreds of years. Overharvest and habitat loss or degradation were

NORTHERN RIVER OTTER *Lontra canadensis*	
COMMON NAMES	American otter, Canadian otter, Fish otter, Nearctic river otter
HEAD AND BODY LENGTH	24–32 inches (61–81 cm)
TAIL LENGTH	12–20 inches (30–51 cm)
BODY WEIGHT	11–31 pounds (5–14 kg)

responsible for the extirpation of many populations in the twentieth century. Because they are one of the top carnivores in aquatic ecosystems, river otters also are very susceptible to the bioaccumulation of toxic chemicals like mercury and PCBs in the food web. Obviously, available habitat is necessary to maintain river otters. Draining marshes, stream channelization, and other wetland losses negatively affect populations. Once extirpated in parts of the United States, river otters have been successfully reintroduced in numerous states. Today, they are becoming more common throughout much of their former range, and occur in national parks from Kobuk Valley to the Everglades, and Acadia to the Grand Canyon. In some areas, however, they remain rare or extirpated. Often blamed for reducing game fish populations, otters very rarely take trout or other popular sport species. River otters sometimes cause economic problems, however, by taking fish at hatcheries and commercial ponds. Where populations can sustain trapping pressure, river otters remain a valuable fur resource, and total annual harvest throughout North America is about twenty-five thousand individuals.

Black-footed Ferret

Probably the most endangered mammalian species in the United States, the black-footed ferret has been a victim of its highly specific feeding regime. Because it preys almost entirely on prairie dogs and uses their burrows for shelter, this narrow specialization of ferrets almost led to its extinction. Black-footed ferrets were ancillary victims of the ill-advised and intentional eradication of about 98 percent of the prairie dog colonies. Tens of millions of prairie dogs were poisoned and countless square miles of colonies destroyed to clear land for agriculture and livestock. Historically, black-footed ferrets were fairly common and enjoyed a widespread distribution throughout the grasslands from southern Alberta and Saskatchewan and the central United States to Texas and Arizona, and south to Mexico. Today, through captive breeding programs and reintroductions, they have returned to several sites throughout their former range.

Black-footed ferrets are large weasels, about the size of a mink. Males are about 10 percent larger and heavier than females. Like all weasels, black-footed ferrets have long, thin bodies with relatively short legs—the perfect shape to enter prairie dog tunnels. Pelage color is a yellowish-brown, and the tip of the tail, legs, and feet are black (the species name

is Latin for "black foot"). Black-footed ferrets also have a distinctive black facemask. The term ferret is derived from the French *fuiret* for "thief"—a reflection of their mask.

Tracks or evidence of digging can be seen, but it is unlikely that most people will observe black-footed ferrets, and not just because of their rarity. They are nocturnal and spend the vast majority of their time underground in prairie dog tunnels. Using a light from a vehicle, people are most likely to see black-footed ferrets above ground for a few minutes at night as they move among burrow openings. Although greater than 90 percent of their diet consists of prairie dogs, ferrets also opportunistically take other prey associated with the colonies, such as mice and rabbits. Ferrets are rarely above ground because they are prey for coyotes, foxes, badgers, and rattlesnakes.

Black-footed ferrets breed from March through April. A litter of three to four young (kits) is born following a gestation of about forty-three days. Unlike most other members of the weasel family (Mustelidae), black-footed ferrets do not exhibit delayed implantation. Kits emerge from dens at about six weeks of age, by which time they are nearly full grown, and disperse during autumn. They are sexually mature by nine months of age and are solitary except during breeding.

CONSERVATION CONCERNS
The ultimate survival of this species is certainly open to question. Thought to be extinct, a small population of about a hundred black-footed ferrets was discovered in Meeteetse, Wyoming (near Yellowstone National Park) in 1981. Most of these individuals died of canine distemper a few years later, and the eighteen survivors were live trapped from the wild and placed in a breeding facility. By 1991, there were more than three hundred black-footed ferrets in several zoos and breeding facilities, and reintroductions to the wild began. Black-footed ferrets have been relocated to sites in Montana, Wyoming, and Arizona. In 1994, thirty-six were released in Badlands National Park

The long, thin body of this black-footed ferret is ideal for moving through narrow tunnels in search of prairie dogs. One of the most endangered mammalian species in North America, the black-footed ferret spends very little time above ground.

BLACK-FOOTED FERRET		
Mustela nigripes		
HEAD AND BODY LENGTH	15–17 inches (38–43 cm)	
TAIL LENGTH	4–5.5 inches (10–14 cm)	
BODY WEIGHT	2–2.4 pounds (0.9–1.1 kg) males	
	1.3–2 pounds (0.6–0.9 kg) females	

in southwestern South Dakota, where natural reproduction has since been documented. Because of their highly specialized life history and resulting narrow margin of ecological amplitude, the future of these charismatic animals remains uncertain. Maintenance of prairie dog colonies large enough to support black-footed ferret populations is critical for continued recovery. Disease and factors such as inbreeding depression may ultimately drive populations to extinction. Wildlife managers ultimately hope to establish fifteen hundred individuals among ten sites, and it is encouraging that today the species is in much less peril than it was twenty years ago, although the black-footed ferret remains our most endangered mammal.

Mink

Mink are among the many diverse species in the weasel family (Mustelidae). People may not know all the members of this varied family, but they usually are aware of the soft, thick, and glossy fur of mink coats and other apparel—a symbol of luxury for generations. Pelage of wild mink is a dark chocolate brown, often with splashes of white on the chin and chest. Mink are large weasels, with males about 10 to 20 percent larger and heavier than females. Like all weasels, mink have an elongated body with relatively short, sturdy legs, and a tail that is somewhat bushy at the tip. Semi-aquatic, with water-repellent fur, mink also have partially webbed feet. They are much larger, stouter, and darker than long-tailed, short-tailed, or least weasels. Unlike these other weasels, the pelage of mink is not white during the winter.

Mink enjoy a widespread geographic distribution in North America throughout most of Alaska, Canada, and the United States, except for Arizona and arid regions of the Southwest. They usually are common in streams, lakes, marshes, or riparian areas where there is permanent water. Dens are in downed logs, rocks, brushy areas, or abandoned muskrat houses along banks or shores. Mink are solitary, primarily nocturnal, and active throughout the year. Muskrats are a favored prey item, but mink feed opportunistically on mice, rabbits, waterfowl, crayfish, frogs, and slow-moving fish such as carp. They are adept swimmers, can dive to depths of twenty feet, and can travel submerged for one hundred feet. Mink cache excess food, and stockpiles of carcasses can be quite large. Like all mustelids, mink have anal glands that produce an acrid musk as strong and pungent as that of skunks. Musk is deposited on rocks and stumps to delimit defended territory, and fights between mink are common when an individual trespasses into an occupied area.

Although there are exceptions (including black-footed ferrets), most mustelids delay implantation of fertilized eggs in the uterus. Mink are variable, however, depending on temperature and light conditions. Females that breed early in the season may delay implantation, but those that breed later probably do not. Thus, time from mating to birth can range from forty to seventy-five days, but actual gestation is only about thirty-two days. Breeding occurs in February and March, with neonates born in April or May. Females produce one litter a year with an average of four altricial young (kits). At birth, kits weigh only 0.3 ounce and are about three inches long. They grow quickly, however, and are weaned when five to six weeks old. Young disperse in autumn and are sexually mature when ten months old. Mink can live up to six years in the wild. Their primary predators are coyotes, lynx, bobcats, red foxes, alligators, and great horned owls.

CONSERVATION CONCERNS

One of the premier furbearers in North America, mink have been trapped for centuries for their highly valued pelts. From four to seven hundred thousand wild mink are trapped annually throughout North America. Only about 10 percent of marketed mink furs are from wild animals, however. The other 90

MINK *Mustela vison*	
COMMON NAMES	Least otter, Water weasel
HEAD AND BODY LENGTH	14–19 inches (36–48 cm) males
	12–16 inches (30–41 cm) females
TAIL LENGTH	8 inches (20 cm) males
	7 inches (18 cm) females
BODY WEIGHT	1.5–3.0 pounds (0.7–1.4 kg) males
	1.2–2.5 pounds (0.5–1.1 kg) females

percent are from captive-raised mink on fur farms, where artificial breeding produces such fur colors as amber gold, iris blue, platinum, sapphire, and several other varieties from white to black. There also are numerous fur farms for mink in Europe, Scandinavia, and Russia. Because wild mink are clearly dependent on aquatic habitats, populations are vulnerable to drainage of wetlands, dam construction, stream channeling, and toxic pollutants such as DDT and PCBs that accumulate in aquatic food webs. Despite trapping mortality and habitat loss, mink populations are stable throughout most of their range.

Wolverine

Powerful, gluttonous, bold, pugnacious, and ferocious are among the many adjectives commonly applied to the wolverine. Whether it is folklore or fact that they drive grizzly bears and cougars away from kills, as suggested by some authorities, wolverines are well known for their aggressive and fearless nature. They prey primarily on small mammals and birds but are known to take caribou or moose as well, especially if these ungulates are sick, injured, or trapped in deep snow. Wolverines consume large amounts of carrion, as well as berries and fish when available. Excess food is cached, and the wolverine deposits a foul-smelling musk from the anal glands to "mark" its property.

WOLVERINE *Gulo gulo*	
COMMON NAMES	Glutton, Skunk bear, Devil bear
HEAD AND BODY LENGTH	31–33 inches (79–84 cm) males
	25–28 inches (64–71 cm) females
TAIL LENGTH	7–10 inches (18–25 cm)
BODY WEIGHT	24–40 pounds (11–18 kg) males
	13–26 pounds (6–12 kg) females

Pelage is dark reddish to blackish brown, with a distinctive tan band running down each shoulder and side to the base of the bushy tail. The throat and chest have lighter blotches of fur as well. Guard hairs are long and coarse and overlay thick underfur. This dense pelage allows wolverines to be active regardless of snow and cold temperatures. Their teeth are large and heavy. Like hyenas, dentition is ideally suited for crushing bone and tearing through tough hides. These solitary hunters are the largest terrestrial members of the weasel family (Mustelidae). Males are much larger than females and weigh about 50 percent more.

The geographic range of the wolverine extends throughout Alaska and northern Canada and south to the higher elevations of Pacific Northwest states. They occupy habitats from tundra in the north to conifer forests, old fields, and plains in more southern areas. Both sexes have extensive home ranges of one hundred square miles or more. Within a home range, certain portions may be defended as territories to keep other wolverines out. Primarily terrestrial, wolverines can climb and are excellent swimmers.

Males and females come together for the breeding season, which extends from April to August. Like most other mustelids, wolverines undergo delayed implantation. Young are born the following spring after actual gestation of thirty to forty days. Thus, a mature female has a litter of two to four kits every other year. Newborns are about five inches long and weigh only three to four ounces. They nurse for nine to ten weeks.

CONSERVATION CONCERNS

Population densities are believed to be reduced from historical levels, as is the geographic range of the wolverine. But given its solitary nature and large home ranges, it is very difficult to quantify population dynamics of this species and its general life history is poorly understood. Harvest from hunting and trapping is the greatest mortality factor. About one thousand wolverines are taken annually. Although legally harvested throughout their range in Canada, within the United States they can be taken only in Alaska and Montana. Wolverines are not an important part of the fur market. Nonetheless, pelts are highly valued by natives and trappers as trim on parkas and mittens because they are resistant to buildup of frost. Wolverines are often considered a problem because of their propensity to raid trappers' cabins and food caches. What they don't destroy, they spray with their noxious musk. Wolverines also disrupt trap lines, take other animals that are caught in traps, and even take the bait out of traps. Wolverines are considered

An aggressive, fearless nature is one of the reasons the wolverine is also called a devil bear. Shaggy fur and the typical coloration pattern are evident in this wolverine.

endangered in eastern Canada, as is the subspecies on the Kenai Peninsula of Alaska. Wolverine populations are much more secure in pristine wilderness areas away from human development.

American Badger

The unique head and facial pattern is the "badge" for which the badger is named. Face and head are black with a distinctive white stripe down the middle from the shoulder to the tip of the nose. The ears and cheeks are white, with a black crescent patch on each cheek. In contrast, the rest of the pelage is a less distinctive buff brown with black-and-white-tipped hairs giving it a grizzled appearance. The fur is somewhat coarse and shaggy, especially on the sides. The body is strong, compact, and slightly flattened top to bottom. Badgers are well adapted for extensive burrowing and have short legs with long sharp claws on the forepaws. Males are larger and about 25 percent heavier than females.

American badgers occur from west-central Canada throughout much of the northern, western, and central United States to Mexico. A different species, the Old World badger, once occurred throughout Europe, the Middle East, and Asia. Dachshunds were bred to hunt them, and the term "to badger" derives from the intentional harassment of Old World badgers by dogs.

Badgers occur in fairly dry, treeless habitats ranging from high elevation alpine meadows to deserts, grasslands, and prairies. They are powerful diggers and occur where soils allow for their long, deep burrows. Burrows may be thirty feet long and extend ten feet below the surface. Entrances to burrows are characteristically semicircular. Diggings may be seen much more commonly than the animals themselves. Badgers are primarily nocturnal but may forage anytime. They feed primarily on fossorial species such as pocket gophers, prairie dogs, and ground squirrels. They also take mice and

The unique head and facial pattern of the badger is unmistakable, as are the coarse, grizzled fur, short tail, and large claws of the forefeet.

voles, other small vertebrates including ground-nesting birds and snakes, as well as invertebrates. Extensive areas are excavated as they forage, often creating hazards for livestock and horsemen, and problems for farmers. Badgers are active throughout the year. During severe winter weather, however, they remain below ground for long periods and become lethargic, although they do not hibernate. Predation, primarily by coyotes and golden eagles, is minimal.

Solitary throughout much of the year, badgers come together to breed in late July and early August. Implantation of the blastocyst is delayed until February. Following a gestation of about forty-five days, a typical litter of two young is born in March or early April. Newborns are furred but helpless. They nurse for about six weeks and disperse from the natal area when they are about three months old.

CONSERVATION CONCERNS

Badgers serve a useful purpose in reducing pocket gophers, ground squirrels, and other rodents. Generally, however, they are viewed as a pest because of the damage they do to rangelands and farms with their extensive burrowing. Historically, trapping, shooting, and poisoning eradicated badgers. They are now making somewhat of a comeback in distribution, extending their range east and south throughout the United States. In the past, their fur was used for paint brushes as well as shaving brushes (although these were primarily from European badgers). Today, these products generally are synthetic, and badger fur is of little economic value. Population densities are considered to be stable throughout most of the range, although badgers are protected in a few states.

Skunks

Like porcupines and raccoons, the striped skunk is a mammalian species that is easily recognized by almost everyone, as is its pungent odor. Striped skunks have black pelage with a white head and white fur continuing to a V-shaped stripe or broken pattern along the back. Far from concealing animals, this distinctive pattern is intended as a clear signal to potential predators not to mistake them for something else and attack. The defensive behavior of skunks includes extensive hissing and growling followed by stomping their forefeet on the ground. A skunk will then move into a handstand to make sure an aggressor sees its pelage color and

AMERICAN BADGER	
Taxidea taxus	
COMMON NAMES	Badger, North American badger
HEAD AND BODY LENGTH	20–26 inches (51–66 cm)
TAIL LENGTH	4–5 inches (10–13 cm)
BODY WEIGHT	19 pounds (8.6 kg) males
	14 pounds (6.4 kg) females

pattern—a clear indication of "skunk." While standing on its forefeet, it twists its rump around so both head and anal area are directed toward a potential attacker. At this point, most prudent predators depart, although a certain degree of learning is necessary on the part of predators. If an attack continues (often by naive younger predators), the skunk sprays a musk fluid up to twelve feet. This noxious scent, familiar to most people, is produced from a pair of glands about the size of marbles, at the base of the tail and extruded through two small nipples in the anus. All species of skunks exhibit similar defensive behaviors and can spray.

Striped skunks are found from southern Canada throughout the contiguous United States and into Mexico. They are usually very common in a variety of habitats where they den in rock crevices, downed logs, or burrows. Feeding habits of striped skunks are as variable as the habitats in which they occur. They are true omnivores, taking many insects that are agricultural pests, as well as consuming small animals and plant material. Like all skunks, they are nocturnal. Except for a female with young, they are generally solitary although communal nesting may occur during winter. Striped skunks can breed when they are ten months old, and litter sizes can reach ten young (called kits). Newborns are weaned at about two months of age and disperse soon afterward. Most skunks probably do not live more than three years in the wild.

Until recently, skunks were considered members of the weasel family (Mustelidae). Currently, they are placed in their own family (Mephitidae), with six species recognized in the United States. The striped skunk is the best known and most widely recognized, but there are also eastern and western spotted skunks, hooded skunks, and eastern and western hog-nosed skunks. Striped skunks are larger than spotted skunks but smaller than hog-nosed skunks. As in all species of skunks, males are about 10 percent larger than females. Spotted skunks are the smallest. Their pelage is silkier than that of striped skunks, and their black fur has a distinctive broken pattern of white spots. They have long, bushy tails, and western spotted skunks have white tips to their tails. Eastern spotted skunks range from the midwestern states east to Florida and north to Pennsylvania. Western spotted skunks occur from the West Coast to the plains states. Both species are nocturnal and omnivorous, but they take more animal matter than other skunks. They occupy dens constructed by other species or excavate their own. A major difference between the two species of spotted skunks involves reproduction. Western spot-

ted skunks have a very long delayed implantation. That is, they breed in the fall but the fertilized egg floats free in the uterus for 180 to 200 days before it implants. Once implanted, the actual gestation of about sixty days begins. Eastern spotted skunks have a delay of only about fourteen days prior to implantation and gestation. Striped skunks also have delayed implantation.

Hooded skunks and hog-nosed skunks are much more geographically restricted, and their biology is less well known than that of striped or spotted skunks. Hooded skunks occur primarily in Mexico, reaching their northernmost distribution in southwestern Texas, New Mexico, and southern Arizona. They are found in rocky arid regions where they primarily consume insects. Like all skunks, they are omnivores and take rodents, small birds, and vegetation. Little is known about reproduction in hooded skunks.

Just as western spotted skunks are smaller than their eastern counterpart, western hog-nosed skunks are about 25 percent smaller than eastern hog-nosed skunks. Eastern hog-nosed skunks are the most geographically limited of the North American skunks, restricted to southern Texas, and do not occur on any national park. Western hog-nosed skunks occur throughout southern Texas, Arizona, and New Mexico as far north as Colorado. Both species have a large naked nose pad that gives them their common name. They inhabit arid scrub brushlands where they feed opportunistically on insects, small vertebrates, and vegetation. As in hooded skunks, little is known about reproduction, although gestation is probably about two months in both species of hog-nosed skunks.

A striped skunk stands on a snowbank alongside a river in Yellowstone National Park. Although it does not hibernate during the winter, in northern areas the striped skunk is inactive and lives off stored body fat.

CONSERVATION CONCERNS

Although most people commonly consider bats to be a significant carrier of rabies, in reality striped skunks are much more likely to be infected (as are raccoons and foxes). Occurrence of rabies in skunks varies considerably from population to population. Each year, however, skunks account for 20 to 30 percent of all cases of rabies reported in the United States and Canada. As a result, management has often involved control programs. On the positive side, all species of skunks are valuable allies in helping to reduce rodents around farms and rural areas, as well as reducing many agricultural insect pests. Many skunks are trapped or shot for their fur, although pelts of all species are economically insignificant, especially the short, coarse fur of hog-nosed skunks. Striped skunks are abundant throughout their range and are of little concern to wildlife conservation programs, aside from the rabies issue. Both species of spotted skunks are wide ranging. However, they often are uncommon, and population status is poorly known, as is that of the hooded skunk. Hog-nosed skunks are the least known of the group. In fact, eastern and western hog-nosed skunks actually may be the same species.

Striped Skunk *Mephitis mephitis*	
COMMON NAMES	Large skunk, Polecat
HEAD AND BODY LENGTH	20–30 inches (51–76 cm)
TAIL LENGTH	7–16 inches (18–41 cm)
BODY WEIGHT	4.5–10 pounds (2–4.5 kg)

Hooded Skunk *Mephitis macroura*	
COMMON NAMES	Southern skunk, White-sided skunk
HEAD AND BODY LENGTH	9–15 inches (23–38 cm)
TAIL LENGTH	14–16 inches (36–41 cm)
BODY WEIGHT	1.8–2.6 pounds (0.8–1.2 kg)

Eastern Spotted Skunk *Spilogale putorius*	
COMMON NAMES	Civet cat, Little polecat, Four-striped skunk
HEAD AND BODY LENGTH	12 inches (30 cm)
TAIL LENGTH	8 inches (20 cm) males
BODY WEIGHT	0.9 pounds (0.4 kg)

Western Spotted skunk *Spilogale gracilis*	
COMMON NAMES	Civet cat, Polecat
HEAD AND BODY LENGTH	10 inches (25 cm)
TAIL LENGTH	5.5 inches (14 cm)
BODY WEIGHT	1.5 pounds (0.7 kg)

Eastern Hog-nosed Skunk *Conepatus leuconotus*	
COMMON NAMES	White-backed skunk, Rooter skunk
HEAD AND BODY LENGTH	17 inches (43 cm)
TAIL LENGTH	11 inches (28 cm)
BODY WEIGHT	7 pounds (3 kg)

Western Hog-nosed skunk *Conepatus mesoleucus*	
COMMON NAMES	Texan skunk, Badger skunk, White-backed skunk
HEAD AND BODY LENGTH	15 inches (38 cm)
TAIL LENGTH	9 inches (23 cm)
BODY WEIGHT	4.4 pounds (2 kg)

COLLARED PECCARY

Easily mistaken for wild pigs, collared peccaries have long pig-like snouts but smaller, thinner bodies. Also, their hind feet have three toes, whereas pigs have four. Peccaries have fairly straight, short canine teeth that form tusks about two inches long; the tusks on pigs are longer and more curved. Pigs are a nonnative (exotic) species, but the three species of peccaries are native to—and found only in—the Western Hemisphere. The collared peccary is the only one that occurs in the United States. It inhabits southwestern Texas and the southern portions of New Mexico and Arizona. Most collared peccaries are in Mexico and Central and South America. Pelage color varies but generally is a coarse, grizzled grayish brown. They have a light band around the neck and shoulders that gives rise to the common name, and there is a scent gland in the middle of the back.

In the semiarid regions where peccaries occur, they consume large amounts of prickly pear cactus. Diets vary regionally, and may include vegetation such as broadleaf plants, nuts, fruits, and occasionally small animals. Foraging areas often show obvious signs of rooting activity. Park visitors are most likely to see peccaries active in brushy areas at dusk or at night during the summer. The animals are much less active on hot summer days as they seek shade and avoid high temperatures. Peccaries also are known as "javelina" (Spanish for javelin), so named by conquistadors because of how quickly the animals can move through thick underbrush. They are territorial and very social, forming groups of ten to eighteen individuals. During winter, group size is larger than in summer, and the animals may huddle together for warmth. As might be expected for a social species, many distinctive vocalizations for group communi-

cation and intra-specific behaviors have been documented. The most commonly observed behavior pattern is two or more animals grooming each other. It is also common to hear them grunting while they forage. The scent gland on the back also functions in communication or when an individual is alarmed.

Female collared peccaries are able to breed by eight months of age. Males are reproductively active when eleven months old, but the dominant male in a group does most of the breeding. Breeding occurs primarily from November through January, and gestation is about 145 days. Litter size usually is two, and the young are active within a few days after birth. Weaning occurs at about two months of age.

CONSERVATION CONCERNS

Collared peccaries are hunted in Texas, New Mexico, and Arizona. An average of about twenty thousand animals is taken each year, most of them from Texas. In certain areas, if peccaries are overharvested, populations can be slow to recover. However, harvests are regulated and the populations are well monitored and generally secure. Collared peccaries are not hunted on the four national parks where they occur, nor are they hunted on Organ Pipe Cactus National Monument. Conversely, collared peccaries are a game species on Amistad National Recreation Area, where they are hunted during the archery and general deer seasons.

COLLARED PECCARY	
Tayassu tajacu	
COMMON NAMES	Javelina, Peccary
HEAD AND BODY LENGTH	33–37 inches (84–94 cm)
TAIL LENGTH	1 inch (2.5 cm)
BODY WEIGHT	30–66 pounds (14–30 kg)

DEER

Deer include some of the most common, charismatic, and easily recognized large mammals in North America. Five native species occur in several national parks: white-tailed deer, mule deer, elk, moose, and caribou. Deer are in the order Artiodactyla, which means "even-toed," for the two toes on each foot that form the hooves.

Antlers characterize males in the forty-three species throughout the world that make up the deer family (Cervidae). Only one species, the Chinese water deer, does not have antlers. Unlike horns, antlers are made of bone and are branched with several points (called tines) that extend from them. Antlers are cylindrical or flattened (palmate) as in moose. Males shed their antlers every year after the autumn breeding season and regrow them beginning in the spring. A growing antler is covered with tissue that has many nerves and blood vessels. This tissue has very dense, soft dark fur called "velvet." Antler growth is the fastest of any vertebrate tissue, and in early summer antlers can grow an inch a day. Once the antler reaches full size, the velvet sloughs off. Large antlers allow males to establish dominance over other males and attract females for breeding. Antler size is a function of quality nutrition, age, and good genes. Many hunters highly value large trophy-size antlers as well.

Deer are strict herbivores. The variety of vegetation consumed depends on location and seasonal availability. They also are ruminants, which means their stomachs are partitioned into four distinct chambers that digest vegetation. Like all mammals, deer have no enzymes that break down cellulose in the cell walls of plants. Instead, they rely on bacteria and other microscopic, one-celled organisms (protozoans) to do the job. These "microfauna" occur in the first three chambers of the stomach: the rumen, reticulum, and omasum. The abomasum is the fourth chamber, or true stomach, and contains the typical enzymes that further digest food.

By the end of the 1800s, most species of deer throughout North America had been drastically reduced in numbers throughout their ranges because of market hunting (commercial selling of venison) and subsequent overharvest, as well as habitat lost to development. Thanks to conservation initiatives and wildlife management programs, including harvest regulations, habitat restoration, and transporting and releasing deer to previously occupied areas, most species of deer are doing well. Subsistence hunting for all cervids, as well as other ungulates

Similar in appearance to a wild pig, a collared peccary forages on desert vegetation in Saguaro National Park. In the United States, the collared peccary is restricted to arid habitats in parts of Texas, New Mexico, and Arizona.

like mountain goats and mountain sheep, is allowed in Alaska's national parks, except in Glacier Bay and Kenai Fjords. Subsistence harvesting also occurs on those parts of Denali and Katmai that are national preserves. There are no seasons or bag limits in place for subsistence hunters. Deer, elk, moose, and caribou are most often seen at dawn or dusk, and these majestic animals are easily recognized and appreciated by park visitors.

Elk

Large antlers, reddish pelage, and prominent rump patch are all clearly seen in these bull elk. During the rut, they will use their antlers to spar and establish dominance for access to breeding females.

Descriptions of elk usually include adjectives such as stately, regal, and majestic. Elk are indeed magnificent animals, larger than all other species of deer except moose. Size varies considerably depending on subspecies. Pelage coloration varies seasonally, from a reddish brown in the summer to a darker head and neck with paler body in the winter. Bulls have a dark, somewhat shaggy neck mane. A yellowish white rump patch flares when an animal is alarmed. Antlers are large and sweep up and back over the neck. In ma-

ture bulls, antlers may be five feet tall, weigh twenty pounds, and have six long tines or points on a side. Bulls produce their largest antlers in the prime of life, between seven and ten years of age, and they are popular trophies among hunters.

Elk inhabit forests, upper elevation meadows and old fields, and grasslands throughout the western United States. Most are on federal lands such as wildlife refuges, national forests, and national parks.

They are opportunistic grazers on grass, shrubs, and forbs. During the summer when they are at higher elevations, they are most active feeding in the morning or late afternoon. Elk are very gregarious and form herds of several hundred animals, moving from higher elevation areas in the fall to warmer valleys in the winter. Sociality varies considerably depending on the season. Sexes form separate herds or occupy different areas much of the year.

As with other deer, the rut occurs in the fall. It is a raucous combination of antler thrashing, wallowing, and digging accompanied by the unmistakable sound of mature bulls bugling. These loud bellows and whistles allow competing bulls and potential mates to assess each other's strength and stature. Bulls of equal size spar with heads lowered, pushing and shoving with their antlers. Smaller, younger bulls do not attempt to challenge larger dominant males. Bulls and cows can breed as yearlings although three to four years of age is typical. Young males usually have limited opportunity to breed because of older dominant bulls, which maintain breeding groups of up to thirty cows. Gestation is about 240 days and a single young is typical. Newborns (calves) usually weigh about thirty pounds, are reddish brown with light tan spots, and are capable of moving almost immediately. They spend their first week or two lying hidden. Hiding is necessary because predation by wolves, grizzly bears, black bears, cougars, and coyotes is a major mortality factor on calves.

Confusion often surrounds the common name of elk. In North America, elk are also called wapiti, a Shawnee Indian word meaning "white rump." This name may be preferable because elk in Europe are called moose and moose are called elk. Further confusion in common names results from the fact that European red deer are the same species as North American elk. Regardless of what they are called, these "stately, regal, and majestic" animals are highly valued by people and are an integral and easily observed part of the large mammal fauna of many of the national parks.

CONSERVATION CONCERNS

Elk once ranged over much of the United States in forested and grassland habitats. By the early 1900s,

ELK *Cervus elaphus*	
COMMON NAMES	North American elk, Wapiti
ADULT BODY LENGTH	7–8 feet (210–240 cm)
TAIL LENGTH	11–14 inches (28–36 cm)
BODY WEIGHT	440–1,000 pounds (200–454 kg) males 350–800 pounds (160–364 kg) females

they were drastically reduced throughout most of their range because of habitat loss and overharvest. Elk were taken in excess by commercial market hunters and sometimes killed simply to extract the upper canine ("elk tooth") as a souvenir. Although some elk populations have returned naturally, many throughout the United States today are the result of reintroductions, habitat acquisition and management, and strict hunting regulations. Reintroductions have included herds in North Dakota, Arkansas, Michigan, Wisconsin, North Carolina, Pennsylvania, and recently Kentucky. A critical aspect of good habitat management for elk is controlled burning, which retards plant succession. Changes in policies regarding burning at the state and federal levels to promote a more natural fire regime benefit elk habitat. Today, elk have rebounded throughout most of their range. Elk are even raised commercially on private game farms throughout North America to provide breeder stock, venison, and antler velvet. Interestingly, Grand Teton National Park is the only one that allows elk hunting, as part of an early political arrangement.

Mule Deer

The common species of deer throughout western North America, mule deer range from extreme southern Alaska through western Canada and the United States, the Plains states to the central Dakotas, Nebraska, and Kansas, and south to Mexico. Mule deer on the Pacific Coast and in Alaska include a separate group of subspecies known as black-tailed deer. Regardless of subspecies, all mule deer have characteristic large ears up to ten inches long. Their ears are much larger than those of the closely related white-tailed deer, and are the basis of their common name. Several other physical features distinguish mule deer from whitetails. Mule deer are stockier and more barrel chested, with shorter tails that they do not elevate ("flag") when they run. Unlike whitetails, antlers of mule deer are dichotomously branched, which means the main beam splits into two equal branches, and each of these branches again to form

MULE DEER *Odocoileus hemionus*	
COMMON NAMES	Black-tailed deer, Sitka deer
ADULT BODY LENGTH	4–5 feet (122–150 cm)
TAIL LENGTH	5–9 inches (13–23 cm)
BODY WEIGHT	88–264 pounds (40–120 kg) males
	66–176 pounds (30–80 kg) females

points (tines). Mule deer also have an unusual stiff-legged, bounding gait, called stotting. It is fascinating to watch them as all four legs leave and hit the ground at the same time. Although they are not as graceful as whitetails, for short distances mule deer can reach top speeds of twenty-five miles per hour, with bouncing leaps fifteen feet long. Body weights of mule deer vary geographically. The largest males (bucks) are in the Rocky Mountains region, but those along the coast are much smaller. Females (does) usually are about 20 percent smaller in size and weight. Pelage coloration ranges from reddish to tan or dark brown, depending on subspecies and season.

Throughout their range, mule deer occupy deciduous and coniferous forests, brushlands, meadows, prairies, and arid scrub habitats. Individuals migrate between higher elevation summer ranges and lower elevation winter ranges that have less snow cover. Where their geographic range overlaps with that of white-tailed deer in the northwestern and southwestern United States, mule deer usually segregate into different habitats, although the two species are capable of hybridizing. Mule deer are typical of the deer family (Cervidae) in feeding on

a variety of vegetation depending on season and locality. They are most active while feeding at dawn and dusk. Generally, mule deer graze on grasses and herbaceous material in the summer and browse on more woody material during winter, with agricultural crops taken whenever they are available. Acorns are a particularly nutritious food as deer put on fat in preparation for the winter. Mule deer lose about 20 percent of their body weight each winter,

Mule deer occur throughout most of western North America. Their very large ears help distinguish mule deer from closely related white-tailed deer.

which is characteristic for deer.

Also typical of cervids, mule deer breed in the fall. Bucks spar with each other to establish dominance hierarchies and attempt to group females. The term *harem* is not really appropriate for groups of females, because bucks cannot keep does together as a unit. Females move from one breeding group to another. Mule deer are capable of breeding as yearlings, although most young males are prevented from doing so by older, dominant bucks. Gestation is about seven months, the same as white-tailed deer. Young females usually have a single fawn, whereas more mature does produce twins, depending on their body condition. Newborn fawns are spotted and weigh from five to twelve pounds. They remain hidden among vegetation for the first week or two as they nurse. Weaning occurs by about four months of age, by which time they have lost their spotted pelage. Coyotes are the major predators of fawns, and bobcats and golden eagles take a few as well. In certain regions, fawn mortality can have a significant effect on population density. Adult mule deer make up about 80 percent of the diet of cougars.

CONSERVATION CONCERNS

Like all other species of deer in the United States, density and distribution of mule deer were drastically reduced throughout the late 1800s and early 1900s. Primary causes of population decline were habitat loss and overharvest, often through commercial market hunting. Populations have recovered in large part because of conservation and management programs that include habitat acquisition, hunting regulations, and cooperation between state and federal agencies. Early to mid-successional habitats are preferred by most species of deer, including mule deer. Management activities that promote these stages, such as controlled burning, partial thinning of trees, and clearing brush, result in higher densities of deer. Today, sport hunters take mule deer throughout their range and populations generally are stable.

White-tailed Deer

Few people would fail to recognize a white-tailed deer as it bounds away with its long tail raised and waving ("flagging") back and forth. Whitetails occur throughout eastern, midwestern, and Gulf Coast states from Canada to Mexico, including parts of the Pacific Northwest and the Southwest. The species is absent only from California, Nevada, and Utah. As might be anticipated given its broad lati-

tudinal range, body size and weight vary dramatically. The extremely small subspecies of Florida Key deer have body weights of only about fifty pounds; the subspecies found in the Chisos Mountains of Big Bend National Park is very small as well. In contrast, exceptionally large males (bucks) up to 450 pounds occur in northern areas. Females (does) are generally about 20 percent smaller than the bucks. Antlers of whitetails have one main beam with several points (tines) coming off it, unlike the dichotomously branching antlers of closely related mule deer. Antlers begin growing in early spring, when they are covered with a highly vascularized, finely furred tissue called velvet, and reach full size by late summer. As in other deer, antlers are shed following the rut in late autumn or early winter and regrown again the following spring. Antler size in all species of deer is a function of nutrition, genetics, and age (although the number of antler points does *not* equal an individual's age). A mature buck in good condition may have twelve or more points per antler. Besides their antlers, whitetails differ from mule deer by having longer tails, smaller ears, less stocky body shape, and a graceful, bounding gait as they run. Pelage color varies geographically as well as seasonally. It is a reddish brown to tan in summer and lighter gray during winter. Besides visual displays such as tail flagging, whitetails communicate with each other through scent released by glands on the forehead, face, and hind legs, and between the hooves.

Throughout the geographic range, this highly adaptable species occurs in north-temperate woodlands and rainforests, old fields and brushland, and agricultural areas. A mosaic of different habitat types throughout an area is preferred. White-tailed deer are primarily browsers, and feed on a variety of vegetation depending on the season. Leaves, grasses, and forbs are eaten in the fall along with acorns and other mast. During winter, deer reduce their time spent foraging. In northern areas, many individual whitetails may concentrate in small areas. These "yards" are sheltered areas with conifer cover that reduce snow depth and wind. In spring and summer, row crops such as corn and soybeans may suffer heavy damage from foraging deer. Whitetails are most active and easily observed early in the morning or evening.

Does are capable of breeding their first fall when only six months old, but most do so initially as yearlings. Adult females in good condition generally have twins, and in exceptionally good habitats can produce three fawns per litter. Gestation is about two hundred days, and most fawns are dropped in late May or early June. They weigh five to ten pounds and have spotted pelage. Although fawns are capa-

ble of moving soon after birth, they spend much of their first month hidden from predators. They begin to take vegetation by about two months of age, at which time they are weaned. Female fawns may stay with their mother for two years or more, but males generally are forced to disperse from their natal area when they are yearlings. Historically, wolves and cougars were the principal predators of whitetails. Coyotes, bobcats, and black bears can take large numbers of fawns in some areas. In Everglades National Park, for example, bobcats are the primary predators of fawns.

CONSERVATION CONCERNS

One of our most common and widespread ungulates, white-tailed deer provided subsistence for Native Americans and then early European colonists. Indians used all parts of the animal including meat, hides, antlers, internal organs, teeth, and hooves. Loss of habitat and overharvest by settlers seriously reduced population density and distribution of whitetails. It is difficult to realize that white-tailed deer were extirpated throughout much of their original range in the United States by the early 1900s. Many of the places they now occur are the result of reintroductions and successful wildlife management and habitat restoration programs. As a result, over the last thirty to forty years, white-tailed deer have become commonplace and even overabundant in many places, and are again a popular game animal throughout the United States.

Sport hunting is the primary method of keeping populations in check. Whitetails are the most popular big game animal in terms of the number harvested, with recreational hunters taking an average of two million every year. Population densities depend on habitat quality and quantity, but whitetails have adapted well to living in close association with human development. In many places they have exceeded the "cultural carrying capacity"—people may feel there are too many deer. In high densities, whitetails damage crops, seriously impede forest regeneration, and destroy gardens and landscaping. In some highly populated areas, especially where sport hunting is limited, the most common mortality factor for deer is being hit by motor vehicles. Every year, over a million whitetails are hit, with economic losses of more than $1 billion.

The major challenge in most states today is to limit recruitment and population size by removing a sufficient number of mature does through hunting. Where hunting is not feasible, trapping and relocating deer often is the most popular alternative. Unfortunately, there usually are very few unoccupied areas available to release the deer. Disease and starvation ultimately reduce population densities that are well beyond the carrying capacity of the habitat. A recent phenomenon in deer, chronic wasting disease, similar to mad cow disease, is a major concern in states where it has been documented and in surrounding areas. It could have a significant impact on population regulation and management activities of white-tailed deer for many years to come.

Moose

With their long legs, prominent hump between the shoulders, and long bulbous nose, moose are magnificent, if somewhat ungainly, animals. They are the largest living member of the deer family (Cervidae). Adult males (bulls) of the largest subspecies reach body weights up to seventeen hundred pounds. Their huge, flattened (palmate) antlers can weigh up to seventy-five pounds and reach five feet across. As with all deer, their antlers are shed in the autumn or early winter following the breeding season (rut) and regrown each spring. Antlers are largest when a bull is in the prime of life, between seven and eleven years of age. Moose also have a prominent flap of skin, called the dewlap, or "bell," that hangs below the throat, the function of which is unknown.

Moose are geographically restricted to colder, northern areas, including Alaska, Canada, and the northern tier states south through the Rocky

The white-tailed deer is one of the most easily recognized large mammals in the United States. Whitetails are a popular game animal among hunters, but in many parts of their range are overabundant and damage habitat.

White-tailed Deer *Odocoileus virginianus*	
COMMON NAMES	Several depending on subspecies
ADULT BODY LENGTH	2.5–7.5 feet (0.8–2.3 m)
TAIL LENGTH	4–14 inches (10–36 cm)
BODY WEIGHT	55–330 pounds (25–150 kg)

Adult bull moose like this one have immense flattened antlers that are shed every year following the breeding season. The moose is the largest living member of the deer family.

Mountains. They occupy various habitats but usually are associated with coniferous forests of spruce, pine, and fir. Moose frequent recently disturbed or burned areas where forage is most nutritious, and often are around ponds or bogs. They often are seen in water and are excellent swimmers. Because of their large size, they can have difficulty dissipating heat, which probably limits their southern distribution. Cold winter weather is not a problem for moose, given their large size and long, thick pelage. But prolonged periods of deep snow can cause moose to expend a great deal of energy to move from place to place, and snow covers forage that otherwise would be accessible for feeding. When snow cover develops a crust over the top, it poses additional problems. Moose may break through because of their size and weight. Predators such as wolves, however, may be able to move across the crusted snow, increasing chances of predation. Besides wolves, grizzly and black bears are significant predators of moose.

Like other members of the deer family, moose

Moose *Alces alces*	
COMMON NAMES	Elk (in Europe)
ADULT BODY LENGTH	9 feet (3 m)
TAIL LENGTH	4 inches (10 cm)
BODY WEIGHT	770–1320 pounds (350–600 kg) males
	595–880 pounds (270–400 kg) females

are susceptible to a roundworm parasite carried by white-tailed deer. It infects the tissues of the brain and causes a fatal neurological illness. Referred to variously as "moose sickness," meningeal disease, or "whirling" disease, in the final stages moose are debilitated and turn in circles before they die or are killed by predators. In many parts of their range outside of Alaska, fatalities also can occur from huge infestations of winter ticks that cause loss of blood and hair.

Like all deer, moose are ruminants and consume hundreds of species of woody plants. Diets vary seasonally and regionally depending on plant quality and availability, but willow, alder, and birch are commonly taken. A large moose population in an area can have a significant impact on habitat because individuals may consume up to forty pounds of twigs, leaves, and shoots a day. In spring and summer, moose are often near water and take aquatic vegetation where available. During the winter when food is scarce, moose in the eastern part of their geographic range browse on conifers.

Cow moose generally mate by one and a half to two years of age. They usually have a single calf, although twins are common in herds on a high nutritional plane. Bulls are capable of breeding as yearlings but generally need the increased body weight and antler size that comes with age before they gain access to females. The rut occurs during autumn. Most births are in late May following a gestation of about 230 days. Newborn calves weigh about thirty pounds. Unlike most other species of deer, calves are not spotted but are a solid reddish brown. Moose are not gregarious; a mother and her calf are the strongest social unit and calves stay with their mothers during the first year.

CONSERVATION CONCERNS

Moose populations respond well to early successional habitats. As with elk and other deer, controlled burning retards plant succession and can be an important management tool in many areas to help assure adequate forage resources and healthy populations. Likewise, timber cutting and the subsequent growth of understory vegetation increase forage availability for moose. Moose have provided subsistence for indigenous North American cultures for many centuries, and today they continue to provide hunting and other recreational opportunities. More than eighty thousand moose were harvested in North America in 2001, most of them in Canada and Alaska. Generally, however, populations are stable or expanding, and for visitors to the national parks, the sight of a moose is always exciting.

Caribou

In many ways, caribou are unique among the deer family (Cervidae). No other species of deer ranges north of the treeline. This extreme northern distribution influences many aspects of their life history. Caribou also are the most gregarious cervids, forming herds as large as a hundred thousand individuals. They migrate longer distances between summer and winter ranges than any other deer. Finally, caribou are unique in being the only species of deer in which females (cows), as well as males (bulls), have antlers.

Smaller than moose or elk, caribou are only about three feet at the shoulder. Body weights vary significantly both seasonally and geographically, but females generally weigh about 30 percent less than males. Pelage color is gray, tan, or dark brown, with a lighter rump and belly. Hooves are almost round—nearly as wide as they are long—and function like snowshoes. Caribou antlers vary considerably among individuals in shape and size, and often are asymmetrical. Bulls have very large, curved antlers with numerous branching, flattened points, or tines. One of the brow tines that extends over the face is always enlarged and flattened. Although called the "shovel," it is not used for clearing snow. Cows have smaller, less massive antlers than bulls. As in other deer, antlers are used to establish dominance among individuals. Bulls grow their antlers in spring and summer and shed them following the breeding season (rut) in late fall. During the rut, like other deer, bulls use their antlers to spar with each other and establish dominance. Cows grow their antlers later than bulls and consequently shed them later. During the harsh winters when snow-covered forage is limited, cows still have their antlers and can displace antlerless bulls from feeding sites. This certainly benefits cows, many of which are pregnant during this stressful time of reduced resources.

Most cows first become pregnant when they are two and a half years old. A cow in poor condition because of low fat reserves will not breed until the following year, however. A single calf is born in late May or early June after a gestation of about 230 days. Newborns weigh ten to twelve pounds and are able to walk shortly after birth. Unlike many other deer, pelage of calves is not spotted.

Many caribou herds spend considerable time throughout the year on the harsh treeless tundra, which offers few places to hide from predators. In natural ecosystems, their major threat is from wolves. Wolves can take adult caribou, but primarily they prey on a very large percentage of the calf crop each year. For an individual caribou there is safety in numbers. A good strategy to reduce the chance of being attacked is to "get lost" among tens of thousands of other caribou in a huge herd. But large herds need abundant food resources, and forage on the tundra is limited. Because a given area cannot support large herds for long, caribou must move and groups break up and reform throughout the year. Like other deer, caribou eat a variety of plants as they become available seasonally. But unlike other deer, caribou also eat lichens, especially in the fall and winter. Because a blanket of snow covers vegetation much of the year, caribou must dig through the snow with their paws. "Caribou" is an Athabascan Indian word for "digger."

CONSERVATION CONCERNS

The venison, hides, antlers, and bones of wild caribou have been essential to the survival of indigenous people for thousands of years. Currently, subsistence hunters take caribou on many of the Alaskan national parks. Besides subsistence hunting, indigenous people have raised domesticated caribou—usually called reindeer—for their products for centuries. Because of their large herds and migration patterns, wild caribou present several unique management and conservation challenges. Size of all the caribou herds fluctuates dramatically because of many interacting factors, including highly variable and usually severe weather, as well as intensity of hunting by natives and historically liberal harvest regulations. Additionally, the extent of predation rates and calf mortality has not always been appreciated. Debate among state and federal agencies, private landowners, hunters, and conservationists continues over how to best manage, monitor, and conserve caribou herds. Issues in this debate include possible control of wolf populations, effects of energy exploration and development in the High Arctic, and protection of calving grounds and traditional migration routes in the face of increased development.

CARIBOU *Rangifer tarandus*	
COMMON NAMES	Reindeer (domesticated caribou); Barren-ground caribou, Woodland caribou, Peary caribou (subspecies)
ADULT BODY LENGTH	4.5–6 feet (1.4–1.8 m)
TAIL LENGTH	6 inches (15 cm)
BODY WEIGHT	240–650 pounds (109–295 kg) males 132–200 pounds (60–91 kg) females

PRONGHORN

(Pages 188–89) Caribou are the most gregarious member of the deer family and form large herds. They also migrate long distances, often crossing bodies of water, like these caribou swimming across the Kobuk River.

Pronghorns are built for speed. They are medium-sized ungulates with a shoulder height of only three feet, a barrel-shaped body, and long, thin legs that allow a lengthy stride. They are the fastest North American mammals, attaining speeds over forty-two miles per hour. Pelage is reddish brown on the back and sides and white underneath, with a short dark mane on the neck and two white throat bands. Males have distinct black jaw patches. Both sexes have a prominent white rump that flares and enlarges when they are alarmed. Unlike most other ungulates, pronghorns have no dew claws, which are remnants of the second and fifth toes, above each hoof. The most unusual feature of pronghorns, however, is what gives rise to their common name. Males have upright horns that consist of a keratinized black sheath with a sharp, curved tip and a short, forward-directed branch or prong. The sheath is over a bony core as in the true horns found in bison, mountain goats, muskoxen, and other members of the family Bovidae. Unlike bovids, however,

Pronghorn occur only in western North America, where they inhabit open prairies and can use their tremendous speed to outrun potential predators. Pronghorns, first described to science by Lewis and Clark, are the fastest land animal in North America.

pronghorns shed their horns each year following the breeding season and regrow them. Horns may reach twelve inches long in pronghorn bucks, whereas those in does are rarely more than a few inches long and have no prong.

Not only are pronghorns built for speed, they inhabit fairly flat, open deserts and grasslands. Although these habitats do not allow them to hide from potential predators, they allow pronghorns to easily spot danger and, given their speed, facilitate escape. Their large eyes are shaded by long eyelashes, and their excellent eyesight further aids survival. Pronghorns are the sole living member of the family

Antilocapridae and are restricted (endemic) to western North America from Canada through northern Mexico. They occur in fifteen western states, with almost half the pronghorns in the United States in Wyoming. Pronghorns forage on grasses, low-lying herbaceous plants, and small shrubs, especially sagebrush, in areas with limited precipitation. They may migrate one hundred miles or more between summer and winter ranges, and deep snow is a major impediment to movement and finding food.

The autumn breeding season (rut) is only about three weeks long. Males generally defend territories and attempt to attract females, who usually breed at about sixteen months of age. Because of competition from older, more dominant individuals, males generally get little chance to breed until they are about three years old. Gestation is about 250 days and twins are most common. Fawns weigh about eight pounds at birth and spend their first two weeks lying hidden in vegetation, even though they are capable runners when only a few days old. Coyotes, golden eagles, and cougars prey on fawns, although adults can outrun predators.

CONSERVATION CONCERNS

Populations are secure in most of the western United States today. Many pronghorn populations are the result of reintroductions that were necessary following overharvest, primarily through commercial market hunting, and habitat loss in the late 1800s and early 1900s. Populations were at all-time lows by 1920, and pronghorns were extirpated from many parts of their range. Fences associated with domestic livestock are a major problem for many populations today. Pronghorns are very hesitant to jump over fences, preferring instead to crawl under them if possible. Thus, wire mesh fences secured to the ground can limit daily and seasonal movements of pronghorns. In Wyoming, for example, livestock fences, as well as development for natural gas drilling, have fragmented habitats and isolated populations. Nonetheless, pronghorns can be considered a conservation success story today because of federal and state management programs that have brought the species back from near extinction. Today, they are popular game animals throughout their range.

PRONGHORN ANTELOPE	
Antilocapra americana	
COMMON NAMES	Antelope, Prongbuck
ADULT BODY LENGTH	3.7 feet (1.2 m)
TAIL LENGTH	5.5 inches (14 cm)
BODY WEIGHT	92–150 pounds (42–68 kg) males
	79–110 pounds (36–50 kg) females

After nearly disappearing since the days when millions of bison roamed throughout much of Canada and the United States, these bison in Yellowstone National Park are a refreshing sight. Bison once formed a critical part of the wolf-bison-prairie ecosystem that was destroyed in the nineteenth century.

AMERICAN BISON

As the largest terrestrial mammals in North America, American bison are easily recognized and unmistakable. (The European bison is a different species.) Bulls are massive and may reach shoulder heights of six feet and body weights exceeding one ton. Their broad head and forequarters, as well as a large shoulder hump, are covered with long, dark, coarse, shaggy hair. This guard hair overlays shorter woolly underfur. The head and shoulders appear much larger, darker, and heavier than the hindquarters. Horns of bison are black, curved, and point inward. As in all members of the family Bovidae, bison horns are not branched nor are they shed annually as are deer antlers. Length of horns may reach twenty inches in bulls and is smaller in cows.

Historically, wild bison ranged on open prairies, meadows, and semiarid habitats, where they grazed on a variety of grasses, sedges, forbs, and woody vegetation. Bison frequented woodlands in the summer for shade or to escape insects, and in the winter when snow depth prevented feeding in more open areas. Large herds necessitated migration in search of adequate forage, and bison certainly had a major impact on vegetation structure, nutrient cycling, and soil compaction throughout their range.

Bulls are often solitary or form small bachelor herds much of the year, until they join cow-calf groups in the summer. Both sexes are capable of breeding when two to three years old. Young males usually get little opportunity to do so until they are about six years of age, however, because of larger, dominant bulls. Breeding occurs in late summer as bulls establish dominance by butting heads and pushing each other so they can eventually "tend" a small group of cows. Following a gestation of nine to ten months, a cow lies down in a secluded area to deliver one or occasionally two calves. The reddish tan calves weigh forty-five to fifty pounds at birth. They can graze when one to two weeks old, but continue to nurse until they are eight months old. Predation by wolves can be a significant factor in the dynamics of some bison populations, and grizzly bears also may take some individuals.

AMERICAN BISON	
Bison bison	
COMMON NAMES	Plains bison, Wood bison, Buffalo
ADULT BODY LENGTH	6.3–10.5 feet (1.9–3.2 m)
TAIL LENGTH	12–24 inches (30–60 cm)
BODY WEIGHT	1,000–2,000 pounds (454–909 kg) males
	800–1,000 pounds (363–454 kg) females

CONSERVATION CONCERNS

Prior to European settlement of the North American continent, the bison population was estimated at thirty to seventy-five million animals—sixty million is a figure often cited. Bison herds were miles wide and stretched to the horizon. The geographic range extended from Alaska through central Canada, throughout most of the United States, and into Mexico. By the late 1800s, this astonishing number of bison was practically extirpated in the United States. Due in part to the commercial market in meat and hides, eradication also was part of a deliberate campaign by the United States government to eliminate the great herds. Eradication of herds was part of the effort to subdue the Plains Indians, who depended on bison for all aspects of their subsistence.

From tens of millions of bison, by 1887 fewer than six hundred individuals remained. The need to save bison was eventually recognized, and refuges and other preserves were established beginning in the early 1900s. Once on the verge of extinction, the return of bison can be considered a conservation success story. By 2002, there were approximately 450,000 bison in the United States. However, 96 percent of these are in fenced, privately held herds on ranches. The days when millions of wild bison roamed free across the landscape are certainly over.

Because bison historically formed large herds, individuals were in close proximity, which perpetuated spread of parasites and diseases. Today, there is concern on the part of cattle ranchers that diseases of bison such as brucellosis, bovine tuberculosis, and anthrax can spread to domestic livestock. Many authorities suggest, however, that disease transmission from bison to cattle is improbable under field conditions. Nonetheless, the possibility has given rise to management regulations concerning movements of wild bison. For example, those that leave the boundaries of Yellowstone National Park during the summer are shot. Bison present other conservation and management issues.

There are questions regarding the validity of the subspecific status of the wood bison, which forms a remnant population in Canada. It is currently considered endangered and is managed as a distinct entity from the more common plains bison. Also, bison can interbreed with domestic cattle (genus *Bos*). Hybrids of cattle and bison have been raised on ranches for centuries because they do better on poor forage and in harsh weather. Several authorities contend bison should be placed in the genus *Bos*.

MUSKOX

The muskox is a unique species in the High Arctic region of both the Old and New Worlds, where it endures long, cold, dry winters and short, cool summers. In North America, the muskox occurs on the tundra of Canada's Arctic islands and mainland, and in coastal parts of Alaska. It has also been introduced to Nunivak Island and Nelson Island off the southwest coast of Alaska.

The muskox is actually smaller than it appears. Only about four feet tall at the shoulder, it has a very long, coarse, shaggy coat and conspicuous shoulder hump that make it look stocky and more massive than it is. Large round hooves with sharp edges provide traction on ice and snow. Coat color is dark brown to black with an off-white to yellowish patch ("saddle") in the middle of the back. During winter, the long outer coat, up to two feet long and almost touching the ground, overlays a very thick, fine underfur (called *qiviut*). Males (bulls) and females (cows) have large dark horns. A prominent rounded portion (the "boss") almost meets in the middle of the forehead then extends out and down each side of the head, before each horn curls forward and out to a sharp point.

Muskoxen browse on grasses, sedges, forbs, and low herbaceous vegetation. During winter, availability of forage is limited by the extent of snowfall and icing, and muskoxen prefer slopes or ridges where the wind helps keep vegetation clear of snow. They are more likely to be in valleys, meadows, or riparian areas during summer. Unlike caribou, they do not exhibit long seasonal migrations. Muskoxen are highly social, with mixed groups of adult females and subadults found throughout the year. Bulls are solitary in the summer but during the winter join groups that may total a hundred individuals. Major predators are grizzly bears and wolves, and muskoxen are well known for their group defensive behavior. When threatened, mature bulls form a circle or semicircle, with their large horns facing outward toward the danger. Younger animals remain behind the adults.

Breeding occurs in August and September, during which mature bulls attempt to mate with several females. Competing bulls engage in threat displays, vocalizations, and scent marking. Establishing dominance also involves two bulls charging each other at high speed and butting heads, the large horn bosses helping to absorb the tremendous force generated. Calves are dropped in late April or early May after a gestation of about eight months. Females usu-

ally have a single precocial calf, which weighs about twenty-five pounds at birth. Calves begin eating grass soon after birth, but continue to nurse until they are twelve to eighteen months old.

CONSERVATION CONCERNS

The defensive ring formed by muskoxen is effective when they are threatened by wolves or other natural predators, but unfortunately makes herds highly vulnerable to hunters with firearms. Trappers, whalers, polar explorers, and natives soon overharvested muskoxen for meat and hides. Muskoxen in Alaska were extirpated by the late 1800s, and much reduced throughout Canada by the early 1900s. Several reintroductions of muskoxen were made from herds in Greenland. Canada regulated harvests in 1917, and established the Thelon Game Sanctuary as a preserve for muskoxen ten years later. President Herbert Hoover established Nunivak Island as a National Wildlife Refuge for the species in 1927. Since then, muskoxen have been successfully reintroduced to other areas in northern and western Alaska.

In Alaska, federal agencies manage muskoxen that occur on national parks and refuges; otherwise, state agencies have jurisdiction. Muskoxen are hunted through an annual quota system in Canada and Alaska. Conservative numbers are harvested, with quotas based on three to five percent of the estimated population sizes. Probably of more com-mercial importance is qiviut, the fine wool underfur of muskoxen. Comparable to cashmere, it is used in very popular wearing apparel and other goods. After near extermination in North America, muskoxen today can be considered a conservation success story, as populations are increasing in numbers and expanding their range into formerly occupied areas.

MUSKOX	
Ovibos moschatus	
COMMON NAMES	Oomingmak
ADULT BODY LENGTH	7.6 feet (2.3 m) males
	6.9 feet (2.1 m) females
TAIL LENGTH	5.5 inches (14 cm) males
	4.0 inches (10 cm) females
BODY WEIGHT	418–880 pounds (190–400 kg) males
	352–418 pounds (160–190 kg) females

MOUNTAIN GOAT

No other ungulates occur at higher elevations above the treeline, or occupy such inhospitable terrain, as mountain goats. Their ability to climb precipitous, rugged slopes and negotiate narrow ledges—often in high winds, extreme cold, and blinding snow—is unparalleled. As suggested by their common name, they inhabit mountains at elevations of twelve thousand feet or more and extremely steep slopes up to 60°. The common name is something of a misnomer, however. Although they certainly occur in mountains, the species is not a true goat. They are most closely related to mountain antelope such as the Japanese serow, Asian goral, and European chamois.

Both sexes have somewhat chunky bodies with shaggy white or yellowish pelage, thick wool underfur, and a beard up to five inches long in older animals. For grasping and extra traction to secure footholds under steep, wet, and often icy conditions, mountain goats have especially flexible large hooves with rough, textured footpads. Their forequarters are very strong and they can climb at a remarkable pace. Sexes can be difficult to distinguish in the field, although adult males (billies) are 10 to 30 percent larger and up to 50 percent heavier than females (nannies). Sharp, tapering black horns curve slightly toward the back. Horns are nine to twelve inches

No other large mammal can negotiate the high elevation, rocky, and often ice-covered precipices traveled by mountain goats. Mountain goats occur on several national parks in Alaska and elsewhere.

or form small groups of a few individuals. Larger groups are composed of females with kids, yearlings, and two-year-olds. Natural salt licks may serve to attract larger congregations of goats. Regardless, groups come together during the breeding season, when males and females are aggressive and use their horns to establish dominance. Unlike other bovids such as bighorn sheep or bison, mountain goats do not butt heads. Instead, opponents stand side to side in opposite directions and circle each other, spearing the flanks with their horns. The mountain goat has thickened areas of hide on its rump, called dermal shields, that help protect it from an opponent's jabs, but serious injuries can occur. Peak rutting activity is in mid-November. Gestation is about 190 days, and most kids are born the last two weeks of May. Nannies usually have one kid although about 25 percent of litters are twins. Newborn kids weigh five to seven pounds and are able to move about within a few days. Kids are weaned at four months of age although they may continue to nurse occasionally. Nannies usually have their first kid at three years of age.

CONSERVATION CONCERNS

Given their harsh, often inaccessible habitats, mountain goats have been less affected than other ungulates by human activities, habitat loss, and commercial market hunting. Today, mountain goats are hunted throughout most of their range as a trophy species on a strict permit basis, meaning a small percentage of any population is taken. Nonetheless, mountain goat populations appear to be very sensitive to overharvest. Hunting pressure can reduce or eliminate local populations, possibly because of fairly low recruitment rates. Wildlife managers may restrict hunters to taking males, but this usually is not practical because, as noted, distinguishing the sexes in the field is very difficult. Mountain goats have been introduced to many areas both in and out of their native range, including Olympic and Yellowstone National Parks. Interestingly, introduced populations of mountain goats seem to increase in density more rapidly and sustain higher harvest pressure than native populations, probably

long in adult males and are slightly shorter and thinner in females. The native geographic range extends from alpine mountain ranges of southeastern Alaska through the Yukon, British Columbia, and Alberta, to Washington, Montana, and Idaho. Populations of mountain goats have been introduced into Oregon, Nevada, Utah, Colorado, and South Dakota.

Mountain goats are most active in the early morning or evening, but activity varies seasonally and with weather conditions. They forage in alpine meadows and on vegetation found on otherwise barren slopes. Goats browse on various grasses, forbs, and herbaceous vegetation throughout the year, and take mosses, lichens, and parts of conifer trees during the winter. Higher elevation alpine sites are used in the summer. Goats migrate to somewhat lower elevations during the winter. They often use windswept areas with steep slopes that are relatively free of snow. At the northern extremes of their range, mountain goats occur more often at lower elevations, for example, near sea level in Kenai Fjords National Park. Remaining at high elevations and on precipitous slopes, however, helps reduce the chance of predation from cougars, their primary predator, as well as grizzly bears and wolves. Coyotes and lynx are secondary predators, and wolverines and golden eagles occasionally take kids. Despite being very sure-footed and agile, goats sometimes die from falls.

For much of the year, males are generally solitary

MOUNTAIN GOAT	
Oreamnos americanus	
COMMON NAMES	Rocky Mountain goat, Snow goat, White goat
HEAD AND BODY LENGTH	3.9–5.2 feet (1.2–1.6 m)
TAIL LENGTH	4–8 inches (10–20 cm)
BODY WEIGHT	210–250 pounds (95–114 kg) males
	132–165 pounds (60–75 kg) females

because good forage is available and there are few predators. In some places, however, introduced mountain goats can cause problems because they overgraze and disrupt fragile alpine plant communities. For example, in Olympic National Park it was necessary to capture and remove more than four hundred mountain goats throughout the 1980s, and management of the species remains a controversial and contentious issue. Although they certainly have habituated to road traffic in some places, including Glacier National Park, increased human activity in other parts of their range may negatively affect mountain goats, a species about which much remains to be learned.

MOUNTAIN SHEEP

Mountain sheep are best known for the behavior of males during the breeding season. They establish dominance by rearing back on their hind legs and charging forward to butt heads, horn to horn. The sound produced from these collisions can be heard for miles around. To absorb the shock of the tremendous force generated from "horn clashing," mountain sheep have heavy, reinforced skulls and the largest horns relative to body size of any mammal. Nonetheless, combatants sometimes suffer injury and death from butting heads. There are actually two species of mountain sheep. Bighorns have heavier, stockier bodies than the somewhat smaller Dall's sheep. Mountain sheep have robust bodies, with adult males (rams) about 30 percent larger and heavier than females (ewes). In both species, the horns are broad and flat on the upper surface so that they can squarely meet an opponent's horns. They are ridged to provide additional traction. Horns of males curve back and curl around the side of the face. Bighorn sheep have large, massive, and tightly curled horns. Dall's sheep have thinner horns that are much more flared from the head; the tips can be three feet apart—twice as far as in bighorns. Females have shorter, more slender horns that curve back but without the curl. Horns of mountain sheep grow throughout life. An individual's age can be determined from the number of growth rings, or annuli, visible on horns. After about eight years of age, however, the tips, which are the oldest part, begin to wear off (a process called "brooming").

There are several different subspecies of both bighorns and Dall's sheep. They vary in size and pelage color, and give rise to several different common names. The short pelage of bighorns is dark brown in the north to a pale gray in desert populations, with a white-gray muzzle, belly, and white rump patch. Dall's sheep are generally white, but depending on subspecies may grade from silver-gray to black.

Bighorn sheep occur from southern British Columbia and Alberta, Canada, through the mountains, foothills, and low elevation deserts of the western United States and Mexico. Dall's sheep occur in the pristine mountainous regions of Alaska, south to northern British Columbia. Both species of mountain sheep prefer open alpine meadows with grasses and shrubs very close to steep, rocky topography that can be used to escape predators. Because habitat requirements are fairly specific, populations are restricted to suitable areas and are not uniformly distributed throughout the geographic ranges. Many populations are small, with fewer than a hundred individuals. Mountain sheep graze on a variety of grasses and forbs during the morning and early evening. Foraging is opportunistic and depends on location, season, and availability. Mountain sheep migrate each year between summer and winter ranges. They move to lower elevations during the winter to escape heavy snow cover and return to progressively higher elevations the next summer as new vegetation emerges. Movements may be limited or extensive, with some males moving thirty miles within their range for food and water, and to avoid predators. Wolves, cougars, grizzly bears, lynx, and coyotes are among the many potential mammalian predators, and golden eagles can take young lambs. Bighorns are particularly vulnerable to lungworm infection, and pneumonia can significantly reduce population numbers.

Bighorns and Dall's sheep are highly gregarious and may form large herds. Within a population, adult males form bachelor groups that remain apart from maternity groups of adult females, sub-adults, and lambs. When they are a few years old, males leave these maternity groups and join a bachelor group. The groups come together during the annual rut, which in northern populations takes place in late November and early December. Larger, dominant, six- to eight-year-old rams do most of the breeding. Females usually have a single lamb in late May or early June, after a gestation of 170 to 175 days. Ewes normally have their first lamb at three years of age, and seek isolated, rugged terrain to give birth. Newborns weigh about eight pounds and are mobile within a day or two. Lambs nurse for about five months and are weaned when they weigh around sixty-five pounds. It is especially critical for juvenile Dall's sheep to gain sufficient weight to make it through their first winter.

A herd of adult females and young Dall's sheep in Denali National Park takes advantage of abundant forage. Dall's sheep, also called thinhorn sheep, inhabit more remote areas than the closely related bighorn sheep.

More southerly desert bighorn populations may breed throughout most of the year.

CONSERVATION CONCERNS

Among hunters, bighorn and Dall's sheep are among the most popular big game trophies. Hunting is highly regulated by wildlife managers to reduce harvest pressure, and generally less than 2 percent of any population is taken each year. Harvest is usually restricted to older rams based on horn size: an animal must have a three-quarters to full curl to be legal. Because mountain sheep are so valuable as trophies, however, poaching can be a problem. Also, because of low numbers, many populations are prone to dying out (extirpation). Current populations of bighorn sheep throughout the western United States are often the result of reintroductions. Original populations were extirpated because of overhunting, loss of habitat, and competition with domestic livestock, which can transmit diseases to the sheep. Another conservation issue in certain parts of the southwest is the introduction of Barbary sheep. These exotics, found in Carlsbad Caverns and Guadalupe National Parks among other places, have a higher reproductive rate, broader diet, and greater dispersal abilities than the native bighorns and can displace them

from areas of overlap. Likewise, exotic burros also compete with bighorn sheep in areas of overlap, including Grand Canyon and Death Valley National Parks. Dall's sheep inhabit some of the most remote areas in North America, so human development and disturbance are not as severe as for populations of bighorn sheep. Unfortunately, mountain sheep lose some of their wariness in protected areas such as parks and refuges, and may habituate to visitors.

DALL'S SHEEP	
Ovis dalli	
COMMON NAMES	Thinhorn sheep, Stone's sheep, Fannin's sheep
HEAD AND BODY LENGTH	4–5 feet (1.2–1.5 m)
TAIL LENGTH	3–4 inches (8–10 cm)
BODY WEIGHT	180 pounds (82 kg) males
	125 pounds (57 kg) females

BIGHORN SHEEP	
Ovis canadensis	
COMMON NAMES	Desert bighorn, Rocky Mountain bighorn, California bighorn
HEAD AND BODY LENGTH	5–6 feet (1.5–1.8 m)
TAIL LENGTH	3–5 inches (8–13 cm)
BODY WEIGHT	160–260 pounds (73–118 kg) males
	115–200 pounds (52–91 kg) females

EXOTIC MAMMALS

The term *exotic* is synonymous with introduced, alien, and nonnative. These terms are applied to any species that is released, either accidentally or intentionally, into an area outside of its native range. Considering all plants, animals, and even microbes, close to fifty thousand exotic species occur in the United States. Invasion of exotics has become one of the most significant and pressing conservation issues in North America and worldwide. Although a few exotic mammals in the United States were released accidentally, most releases were intentional. Motives for release have included additional game species for sport hunting, furbearers for trapping, species as food sources, aesthetic reasons, or as a means of biological control to reduce populations of a different exotic species. Although some exotic species remain fairly benign in terms of their distributions and ecological impacts, others are widespread and have profound negative effects on native flora and fauna.

National parks vary greatly in the extent to which exotics may be present. Many parks have none. Conversely, almost all the mammals that occur in Haleakala and Hawaii Volcanoes National Parks are introduced. Exotic species there include axis deer, European rabbits, wild goats, a variety of rats and mice, as well as the wild pig, horse, and small Indian mongoose noted below. In fact, the hoary bat is the only species of mammal native to Hawaii.

Some exotic species are small and cryptic; others are easily seen. A large, conspicuous exotic is the Barbary sheep, which competes with desert bighorn sheep where both species occur in southwestern states. Barbary sheep (also called *aoudad*) are found in Carlsbad Caverns, Big Bend, and Guadalupe Mountains National Parks and could easily be mistaken for bighorns. A related species, exotic mouflon sheep, also inhabit Guadalupe Mountains and Hawaii Volcanoes National Parks. Hawaii Volcanoes also has a population of exotic European wild rabbits, as do Haleakala and Channel Islands National Parks. Although there is a question as to whether they ever occurred there naturally, the presence of mountain goats in Olympic National Park has resulted in a great deal of debate and controversy. There is little debate, however, over the damage they do to the vegetation, and to soils through cratering and erosion. Mountain goats are native to North America but probably were introduced to Olympic National Park. Feral domestic goats do similar damage in parks such as Haleakala and Hawaii Volcanoes.

In February 1999, President Bill Clinton issued Executive Order 13112 that in part mandated federal agencies (1) to prevent the introduction of invasive species, (2) to control exotic populations in an environmentally sound manner, and (3) to restore native species and habitats in ecosystems that have been invaded. Full eradication of exotics from an area sometimes may be possible, although in many cases it is not practical.

Wild Pig

No exotic mammal has such an aggressively destructive impact on the environment as wild pigs. They were introduced to the United States more than five hundred years ago by Spanish conquistadors, and today occur in at least twenty-three southern states as well as Hawaii. With an estimated four million animals, they are the most abundant free-ranging exotic ungulate in the United States. Wild pigs annually cause an estimated $800 million damage to crops, pastures, and natural habitats through rooting, trampling, and direct consumption. In many areas, they also compete with native wildlife for forage and can transmit diseases to domestic pigs. On the positive side, and the primary reason for widespread recent introductions, is their popularity as a game species among many hunters. Wild pigs are in several national parks, as well as national wildlife refuges and forests. They have been in Virgin Islands National Park since the early 1700s and in Hawaii even longer. When Polynesians arrived there more than fifteen hundred years ago they brought pigs with them. Another influx of pigs arrived in Hawaii with Captain Cook in 1778. In Haleakala and Hawaii Volcanoes National Parks, management efforts are aimed at preservation and restoration of native flora and fauna harmed by pigs and other exotic mammals. As such, control of pig populations involves trapping, shooting, snaring, and attempting to fence them out of critical areas. Likewise, in places like Great Smoky Mountains National Park, thousands of pigs have been shot,

WILD PIG	
Sus scrofa	
COMMON NAMES	Feral hog, Wild boar, Feral pig
HEAD AND BODY LENGTH	46–76 inches (117–192 cm) males
	53–73 inches (135–185 cm) females
TAIL LENGTH	10 inches (26 cm)
BODY WEIGHT	79–330 pounds (36–150 kg) males
	75–200 pounds (34–91 kg) females

trapped, or removed to protect native resources. To a greater or lesser extent, wherever wild pigs have been introduced, these efforts have become a familiar scenario played out by state and federal conservation and management agencies, as well as the national parks, to help preserve natural ecosystems.

Wild Horse and Burro

Horses were introduced to North America by the Spanish conquistadors, as was the closely related burro, or donkey. Since then, horses that were released intentionally or that escaped from domestication have spread to public lands in ten western states, with the majority in Nevada. Most wild burros now are in parts of California and Arizona. Although commonly referred to as "wild," most free-ranging horses and burros today are more correctly called "feral" because they descended from domesticated animals. Horses are well known to everyone, as are burros. Both certainly evoke images of the rugged pioneer traditions and romance of the Old West. To many wildlife resource managers and ranchers, however, horses and burros represent more controversial conservation issues. They compete with native ungulates and domestic livestock for limited forage and water resources and further degrade often-overgrazed habitats by trampling. Prior to 1971, many horses and burros on western rangelands were routinely captured and killed—often in an inhumane manner. As a result of public outcry over eradication of wild horses and burros, and the efforts of numerous animal protection organizations, the Wild and Free-Roaming Horse and Burro Act was passed in 1971. This federal legislation essentially protects both species and directs management activities of the Bureau of Land Management (BLM). Once they were protected, populations increased dramatically because their removal was end-

ed and there was a lack of natural predators. What to do with the ever-increasing numbers of horses and burros has been a major concern, given that populations can increase by 20 percent annually and double in less than four years. The need to reduce numbers resulted in a novel and successful approach—the Adopt-A-Horse program administered by BLM. Since 1973, more than 122,000 horses and burros have been rounded up and adopted by private individuals, most of them in eastern states. Nonetheless, populations of horses and burros grow faster than they can be adopted, so current research involves the use of immunocontraception to help reduce reproductive rates. In addition, horses have been translocated to other areas in an effort to reduce numbers.

Small Indian Mongoose

One of the more unusual exotics in the national parks is the small Indian mongoose, which is common throughout the Caribbean Islands including Virgin Islands National Park on St. John. It was introduced throughout the mid to late 1800s in an attempt to control another exotic species, the black rat, which seriously damaged sugarcane crops. Mongooses successfully reduced the numbers of rats and the damage to crops. This is small consolation, however, for the tremendous problems posed by the mongoose throughout the Caribbean and in Virgin Islands National Park. Because it is such a successful predator on numerous animals in addition to rats, authorities suggest that the Indian mongoose has been responsible for the endangerment and extirpation of more species of mammals, birds, and reptiles than any other exotic deliberately introduced anywhere else in the world. The Indian mongoose also was introduced to Hawaii in 1883, again as a biological control against rats. As in the Caribbean, mongooses soon turned their attention to other fauna besides rats. They are responsible for the extinction of numerous species of endemic Hawaiian birds where they occur on Haleakala and Hawaii Volcanoes National Parks. Also, like those in the Caribbean, mongooses in Hawaii are considered to be a reservoir for the rabies virus and are the focus of continued control programs.

WILD HORSE *Equus caballus*	
COMMON NAMES	Mustang, Feral horse
HEAD AND BODY LENGTH	9 feet (2.7 m)
TAIL LENGTH	3 feet (90 cm)
BODY WEIGHT	990 pounds (450 kg) males 913 pounds (415 kg) females

BURRO *Equus asinus*	
COMMON NAMES	Donkey, African wild ass
HEAD AND BODY LENGTH	7 feet (2.1 m)
TAIL LENGTH	1.5 feet (45 cm)
BODY WEIGHT	200–250 pounds (91–114 kg)

SMALL INDIAN MONGOOSE *Herpestes auroponcatus*	
HEAD AND BODY LENGTH	9–25 inches (23–64 cm)
TAIL LENGTH	9–19 inches (23–48 cm)
BODY WEIGHT	1–9 pounds (0.5–4.0 kg)

SPECIES DISTRIBUTION

- • species present
- P possible
- i introduced
- r reintroduced
- x extirpated

Shrews

	Virginia opossum	Arctic shrew	Barren ground shrew	Cinereus (masked) shrew	Desert shrew	Dwarf shrew	Inyo shrew	Least shrew	Long-tailed shrew	Marsh shrew	Merriam's shrew	Montane (dusky) shrew	Mount Lyell shrew	Northern short-tailed shrew	Ornate shrew	Pacific shrew	Prairie shrew	Preble's shrew	Pygmy shrew	Smokey shrew	Southeastern shrew	Southern short-tailed shrew	Tiny shrew	Trowbridge's shrew	Tundra shrew	Vagrant shrew	Water shrew
ACADIA				•					P					•					•	•							•
ARCHES					•						•																
BADLANDS						•		•									•										
BIG BEND	P				•																						
BISCAYNE	•							•																			
BLACK CANYON OF THE GUNNISON																											
BRYCE CANYON												•															
CANYONLANDS					•						•																
CAPITOL REEF					•																					P	•
CARLSBAD CAVERNS	P				•																						
CHANNEL ISLANDS																											
CONGAREE	•							•													•	•					
CRATER LAKE																										•	•
CUYAHOGA VALLEY	•			•										•						•							
DEATH VALLEY					•																						
DENALI				•								•							•						•		•
DRY TORTUGAS																											
EVERGLADES	•							•						•													
GATES OF THE ARCTIC			•	•								•													•		
GLACIER				•								•							•							•	•
GLACIER BAY				•																						•	•
GRAND CANYON						•	•				•																
GRAND TETON				•		•																				•	•
GREAT BASIN											P																•
GREAT SAND DUNES												P															•
GREAT SMOKY MOUNTAINS	•			•				•	•										•	•	•						•
GUADALUPE MOUNTAINS				P																							
HALEAKALA																											
HAWAII VOLCANOES																											
HOT SPRINGS	•							•														•					
ISLE ROYALE																											•
JOSHUA TREE					•																						
KATMAI		•		•								•															
KENAI FJORDS				•								•															
KOBUK VALLEY																									•		
LAKE CLARK	P			•								•							•				•		P		P
LASSEN VOLCANIC	•					•												•						•		•	•
MAMMOTH CAVE	•							•						•													
MESA VERDE				•		•						•															
MOUNT RAINIER				•						•														•		•	•
NORTH CASCADES	•			•						•		•												•		•	•
OLYMPIC				•						•		•												•		•	
PETRIFIED FOREST				P																							
REDWOOD	•									•						•								•		•	
ROCKY MOUNTAIN				•		•					•	•							•								•
SAGUARO					•																						
SEQUOIA–KINGS CANYON	i											•	•											•		•	•
SHENANDOAH	•			•				•						•					•	•							
THEODORE ROOSEVELT				•							•																
VIRGIN ISLANDS																											
VOYAGEURS		•		•										•													•
WIND CAVE						P		•			P						•										
WRANGELL–ST. ELIAS				•								•							•						•		•
YELLOWSTONE				•		•						•						P									•
YOSEMITE	i				P							•	•		P									•		•	•
ZION				•							P	•															•

200

American shrew-mole	Broad-footed mole	Coast mole	Eastern mole	Hairy-tailed mole	Star-nosed mole	Townsend's mole	Allen's big-eared bat	Antillean fruit bat	Big brown bat	Big free-tailed bat	Brazilian free-tailed bat	California leaf-nosed bat	California myotis	Cave myotis	Eastern pipistrelle	Eastern small-footed myotis	Evening bat	Fringed myotis	Ghost-faced bat	Gray myotis	Greater bulldog bat	Hoary bat	Indiana bat	Jamaican fruit bat	Keen's myotis	Little brown bat	
				•	•				•						P	P						•				•	ACADIA
							P		P	•												P					ARCHES
									•									•				•				•	BADLANDS
									•	•	•		•	•				•	•			•					BIG BEND
											•																BISCAYNE
									•													•					BLACK CANYON OF THE GUNNISON
									•				•					•								•	BRYCE CANYON
							P		P	•			•					•				P				•	CANYONLANDS
							P		P		•															P	CAPITOL REEF
									•	•	•		•	•				•				•					CARLSBAD CAVERNS
									•		•		•					•				•					CHANNEL ISLANDS
			•		•				•		•				•		•					•				•	CONGAREE
•		•							•																	•	CRATER LAKE
			•	•	•				•						•							•	•			•	CUYAHOGA VALLEY
											•		•					•				•					DEATH VALLEY
																										•	DENALI
																											DRY TORTUGAS
	P										P						P										EVERGLADES
																											GATES OF THE ARCTIC
									•													•				•	GLACIER
																										•	GLACIER BAY
							P		•	•	•	•	•					•				•				P	GRAND CANYON
									•													•				•	GRAND TETON
									•		•		•					•				•				P	GREAT BASIN
									•		P											P				•	GREAT SAND DUNES
			•	•	•				•						•	•						•	•			•	GREAT SMOKY MOUNTAINS
									•	•	•		•	•				•				•					GUADALUPE MOUNTAINS
																						•					HALEAKALA
																						•					HAWAII VOLCANOES
			•						•						•		•					•				•	HOT SPRINGS
									P																	P	ISLE ROYALE
									•		•	•	•					•				•					JOSHUA TREE
																										•	KATMAI
																										•	KENAI FJORDS
																											KOBUK VALLEY
																										•	LAKE CLARK
•	•								•				•									•				•	LASSEN VOLCANIC
			•						•						•	•	•			•		•	•			•	MAMMOTH CAVE
							P				P															P	MESA VERDE
•		•				•			•													•					MOUNT RAINIER
•		•				P			•				•					•				•				•	NORTH CASCADES
•		•				•							•												•	•	OLYMPIC
							•		•	•	•		•									•				P	PETRIFIED FOREST
•	•	•				•			•		P		•					P				•				•	REDWOOD
									•									•				•				•	ROCKY MOUNTAIN
									•		•	•	•	•				•				•					SAGUARO
	•								•		•		•					•				•				•	SEQUOIA–KINGS CANYON
				•					•													•				•	SHENANDOAH
									•													•				•	THEODORE ROOSEVELT
								•													•			•			VIRGIN ISLANDS
					•				P													P				•	VOYAGEURS
									•									•				•				•	WIND CAVE
																										•	WRANGELL–ST.ELIAS
									•																	•	YELLOWSTONE
	•								•		•		•					•				•				•	YOSEMITE
							P		•	•	•		•					•				•				P	ZION

202

Legend:
- • species present
- P possible
- i introduced
- r reintroduced
- x extirpated

Park	Appalachian cottontail	Brush rabbit	Desert (Audubon's) cottontail	Eastern cottontail	Marsh rabbit	Mountain (Nuttall's) cottontail	Pygmy rabbit	Swamp rabbit	Alaskan hare	Antelope jackrabbit	Black-tailed jackrabbit	Snowshoe hare	White-tailed jackrabbit	Allegheny woodrat	Bushy-tailed woodrat	Desert woodrat	Dusky-footed woodrat	Eastern woodrat	Mexican woodrat	Southern Plains woodrat	Stephen's woodrat	White-throated woodrat	Botta's (Valley) pocket gopher	Mountain pocket gopher	Northern pocket gopher
	Rabbits								Hares					Woodrats									Pocket Gophers		
ACADIA												•													
ARCHES			•								•				•	•			•				•		
BADLANDS			•	•							•		•		•										•
BIG BEND			•	•							•								•	•		•	•		
BISCAYNE					•													•							
BLACK CANYON OF THE GUNNISON				•							P		•		•							P			•
BRYCE CANYON			•	•							•		•		•	•									•
CANYONLANDS			•								•				•	•			•				•		
CAPITOL REEF			•	P							•	P	•		P	•							•		P
CARLSBAD CAVERNS			•								•								•	•		•	•		
CHANNEL ISLANDS																									
CONGAREE				•	•													•							
CRATER LAKE				•								•			•										
CUYAHOGA VALLEY				•																					
DEATH VALLEY			•	•							•				•	•									
DENALI												•													
DRY TORTUGAS																									
EVERGLADES				•	•													•							
GATES OF THE ARCTIC									•			•													
GLACIER												•	•		•										•
GLACIER BAY												•													
GRAND CANYON			•	P		•					•				•	•			•		•	•	•		•
GRAND TETON												•	•		•										•
GREAT BASIN			•			•	•				•		P		•	•							•		P
GREAT SAND DUNES			•			•					P		•		•					P					•
GREAT SMOKY MOUNTAINS	•			•																					
GUADALUPE MOUNTAINS			•	•							•								•	•		•	•		
HALEAKALA																									
HAWAII VOLCANOES																									
HOT SPRINGS				•				•										•							
ISLE ROYALE												•													
JOSHUA TREE			•								•					•	•					•	•		
KATMAI									•			•													
KENAI FJORDS												•													
KOBUK VALLEY									•			•													
LAKE CLARK												•													
LASSEN VOLCANIC				•							•	•			•									•	
MAMMOTH CAVE				•										•											
MESA VERDE			•	•							•				•				•			•	•		
MOUNT RAINIER												•			•										•
NORTH CASCADES			i									•			•										P
OLYMPIC												•			•										•
PETRIFIED FOREST			•	•							•											•	•		
REDWOOD		•									•				•		•						•		
ROCKY MOUNTAIN				•								•	•		•				•						•
SAGUARO		•								•	•								•			•	•		
SEQUOIA–KINGS CANYON		•	•								•	•	•		•	P	•						•	•	
SHENANDOAH				•										•											
THEODORE ROOSEVELT			•	•		P						•	•		•										•
VIRGIN ISLANDS																									
VOYAGEURS			P									•													
WIND CAVE			•	•		P					P		•		•										•
WRANGELL–ST.ELIAS												•													
YELLOWSTONE			•			•						•	•		•										•
YOSEMITE		•									•	•	•				•						•	•	
ZION			•			•					•				•	•							•		•

203

	Pocket Gophers (cont'd)				Chipmunks																	Woodchucks/Marmots				
	Plains pocket gopher	Southern pocket gopher	Western pocket gopher	Yellow-faced pocket gopher	Eastern chipmunk	Allen's chipmunk	Alpine chipmunk	Cliff chipmunk	Colorado chipmunk	Gray-footed chipmunk	Least chipmunk	Lodgepole chipmunk	Long-eared chipmunk	Merriam's chipmunk	Panamint chipmunk	Red-tailed chipmunk	Siskiyou chipmunk	Sonoma chipmunk	Townsend's chipmunk	Uinta chipmunk	Yellow-pine chipmunk	Alaska marmot	Hoary marmot	Olympic marmot	Woodchuck	Yellow-bellied marmot
ACADIA					•																				•	
ARCHES									•																	
BADLANDS	•										•															
BIG BEND			•																							
BISCAYNE																										
BLACK CANYON OF THE GUNNISON									•		•															•
BRYCE CANYON								•			•									•						•
CANYONLANDS									•																	
CAPITOL REEF								P	•		•									P						•
CARLSBAD CAVERNS			•							•																
CHANNEL ISLANDS																										
CONGAREE					•																					
CRATER LAKE		•																	•		•					•
CUYAHOGA VALLEY					•																				•	
DEATH VALLEY		•													•											
DENALI																							•			
DRY TORTUGAS																										
EVERGLADES																										
GATES OF THE ARCTIC																						•	•			
GLACIER											•					•					•		•			•
GLACIER BAY																							•			
GRAND CANYON								•	P		•									•						
GRAND TETON											•									•	•					•
GREAT BASIN								•			•									•						
GREAT SAND DUNES									•		•															•
GREAT SMOKY MOUNTAINS					•																				•	
GUADALUPE MOUNTAINS			•							•																
HALEAKALA																										
HAWAII VOLCANOES																										
HOT SPRINGS	•				•																				•	
ISLE ROYALE																										
JOSHUA TREE														•												
KATMAI																							•			
KENAI FJORDS																							•			
KOBUK VALLEY																							•			
LAKE CLARK																							•			
LASSEN VOLCANIC						•					•	•	•	•							•					•
MAMMOTH CAVE					•																				•	
MESA VERDE									•		•															•
MOUNT RAINIER																			•		•		•			
NORTH CASCADES																			•		•		•			
OLYMPIC		•																	•		•			•		
PETRIFIED FOREST									•																	
REDWOOD						•											•	•								
ROCKY MOUNTAIN									P		•									•						•
SAGUARO								•																		
SEQUOIA–KINGS CANYON						•	•		P			•		•						•						•
SHENANDOAH					•																				•	
THEODORE ROOSEVELT											•															
VIRGIN ISLANDS																										
VOYAGEURS					•						•														•	
WIND CAVE											•															•
WRANGELL–ST.ELIAS																							•			
YELLOWSTONE											•									•	•					•
YOSEMITE						•	•				•	•	•	•						P	•					•
ZION								•			•									•						•

species present •
P possible
i introduced
r reintroduced
x extirpated

204

Legend:
- • species present
- P possible
- i introduced
- r reintroduced
- x extirpated

Park	Harris's antelope squirrel	Texas antelope squirrel	White-tailed antelope squirrel	Arctic ground squirrel	Belding's ground squirrel	California ground squirrel	Cascade golden-mantled ground squirrel	Columbian ground squirrel	Franklin's ground squirrel	Golden-mantled ground squirrel	Mexican ground squirrel	Mojave ground squirrel	Richardson's ground squirrel	Rock squirrel	Round-tailed ground squirrel	Spotted ground squirrel	Thirteen-lined ground squirrel	Townsend's ground squirrel	Uinta ground squirrel	Black-tailed prairie dog	Gunnison's prairie dog	Utah prairie dog	White-tailed prairie dog	Eastern gray squirrel	Fox squirrel	Western gray squirrel
ACADIA																								•		
ARCHES			•											•									•			
BADLANDS																	•			•					•	
BIG BEND		•									•			•		•										
BISCAYNE																								•		
BLACK CANYON OF THE GUNNISON										•				•							P					
BRYCE CANYON										•				•								•				
CANYONLANDS			•											•												
CAPITOL REEF			•							•				•								•				
CARLSBAD CAVERNS		•									•			•		•										
CHANNEL ISLANDS																										
CONGAREE																								•	•	
CRATER LAKE						•				•																•
CUYAHOGA VALLEY																								•	•	
DEATH VALLEY			•			•						•			•											
DENALI				•																						
DRY TORTUGAS																										
EVERGLADES																								•	•	
GATES OF THE ARCTIC				•																						
GLACIER								•		•			•				•									
GLACIER BAY																										
GRAND CANYON	•		•							•				•		•					•					
GRAND TETON										•									•							
GREAT BASIN			•							•				•				•								
GREAT SAND DUNES										•				P		•					•					
GREAT SMOKY MOUNTAINS																								•	•	
GUADALUPE MOUNTAINS		•								P				•		•				•						
HALEAKALA																										
HAWAII VOLCANOES																										
HOT SPRINGS																								•	•	
ISLE ROYALE																										
JOSHUA TREE			•			•						•			•											
KATMAI				•																						
KENAI FJORDS				P																						
KOBUK VALLEY				•																						
LAKE CLARK				•																						
LASSEN VOLCANIC					•	•				•																•
MAMMOTH CAVE																								•	•	
MESA VERDE			•							•				•		P										
MOUNT RAINIER										•																
NORTH CASCADES							•	P																		•
OLYMPIC																										
PETRIFIED FOREST			•											•		•					•					
REDWOOD						•				P																•
ROCKY MOUNTAIN										•				P												
SAGUARO	•													•	•											
SEQUOIA–KINGS CANYON					•	•				•																•
SHENANDOAH																								•	•	
THEODORE ROOSEVELT									P								•			•					•	
VIRGIN ISLANDS																										
VOYAGEURS									•								P									
WIND CAVE																	•			•					P	
WRANGELL–ST.ELIAS				•																						
YELLOWSTONE										•									•							
YOSEMITE					•	•				•																•
ZION			•							•				•												

205

Legend:
- • species present
- **P** possible
- *i* introduced
- *r* reintroduced
- *x* extirpated

Park	Tree Squirrels (cont'd) Douglas's squirrel	Red squirrel	Northern flying squirrel	Southern flying squirrel	Mountain beaver	Beaver	Muskrat	Round-tailed muskrat	Porcupine	Toothed Whales Atlantic spotted dolphin	Atlantic white-sided dolphin	Baird's beaked whale	Beluga	Blainville's beaked whale	Bottlenose dolphin	Cuvier's beaked whale	Dall's porpoise	Dwarf sperm whale	False killer whale	Frasier's dolphin	Gervais' beaked whale	Harbor porpoise	Killer whale	Melon-headed whale	Northern right whale dolphin	Pacific white-sided dolphin
ACADIA		•	•	P		r	•		•		•		•									•	•			
ARCHES						•	•		•																	
BADLANDS						•	•		•																	
BIG BEND						•			•																	
BISCAYNE				•											•											
BLACK CANYON OF THE GUNNISON		•				•	P		•																	
BRYCE CANYON		•	•			P	P		•																	
CANYONLANDS						•	•		•																	
CAPITOL REEF		•	•			•	•		•																	
CARLSBAD CAVERNS									•																	
CHANNEL ISLANDS															•		•		•			•	•		•	•
CONGAREE				•			•																			
CRATER LAKE	•		•		•	•	•		•																	
CUYAHOGA VALLEY		•		•		•	•																			
DEATH VALLEY									•																	
DENALI		•	•			•	•		•																	
DRY TORTUGAS															•											
EVERGLADES				•				•							•											
GATES OF THE ARCTIC		•	•			•	•		•																	
GLACIER		•	•			•	•		•																	
GLACIER BAY		•	•			•			•			P					•					•	•			
GRAND CANYON		•				•	P		•																	
GRAND TETON		•	•			•	•		•																	
GREAT BASIN						•	•		•																	
GREAT SAND DUNES		•				•	•		•																	
GREAT SMOKY MOUNTAINS		•	•	•		•	•																			
GUADALUPE MOUNTAINS		P							•																	
HALEAKALA																										
HAWAII VOLCANOES																										
HOT SPRINGS				•		•																				
ISLE ROYALE		•				•	•																			
JOSHUA TREE																										
KATMAI		•				•	•		•				•										•			
KENAI FJORDS		•	•			•			•			P					•					•	•			P
KOBUK VALLEY		•				•	•		•				•													
LAKE CLARK		•				•	•		•				•													
LASSEN VOLCANIC	•		•		•	•	•		•																	
MAMMOTH CAVE				•		r	•																			
MESA VERDE		•							•																	
MOUNT RAINIER	•		•		•	•			•																	
NORTH CASCADES	•	P	•		•	•	•		•																	
OLYMPIC	•		•		•	•	•										•					•	•			•
PETRIFIED FOREST									•																	
REDWOOD	•		•		•	•	•		•			•				•	•		•			•	•			•
ROCKY MOUNTAIN		•				•	•		•																	
SAGUARO																										
SEQUOIA-KINGS CANYON	•		•		•	•	P		•																	
SHENANDOAH		•		•		•	•																			
THEODORE ROOSEVELT						•	•		•																	
VIRGIN ISLANDS										•				P	•	•		P	P	P	P		P	P		
VOYAGEURS		•	•			•	•		•																	
WIND CAVE		•	P			P	•		•																	
WRANGELL–ST.ELIAS		•	•			•	•		•							•	•					•	•			
YELLOWSTONE		•	•			•	•		•																	
YOSEMITE	•		•		•	•	•		•																	
ZION		•	•			•	•		•																	

206

Legend:
- • species present
- **P** possible
- **i** introduced
- **r** reintroduced
- **x** extirpated

Species columns are grouped as: **Toothed Whales (cont'd)** — Pantropical spotted dolphin, Pygmy killer whale, Pygmy sperm whale, Risso's dolphin, Rough-toothed dolphin, Short-beaked common dolphin, Short-finned pilot whale, Sperm whale, Spinner dolphin, Striped dolphin, Stejneger's beaked whale, White-beaked dolphin; **Baleen Whales** — Blue whale, Bowhead whale, Bryde's whale, Fin whale, Gray whale, Humpback whale, Minke whale, Northern right whale, Sei whale, West Indian manatee; **Eared Seals** — Guadalupe fur seal, Northern fur seal, California sea lion, Northern sea lion.

Park	Pantropical spotted dolphin	Pygmy killer whale	Pygmy sperm whale	Risso's dolphin	Rough-toothed dolphin	Short-beaked common dolphin	Short-finned pilot whale	Sperm whale	Spinner dolphin	Striped dolphin	Stejneger's beaked whale	White-beaked dolphin	Blue whale	Bowhead whale	Bryde's whale	Fin whale	Gray whale	Humpback whale	Minke whale	Northern right whale	Sei whale	West Indian manatee	Guadalupe fur seal	Northern fur seal	California sea lion	Northern sea lion
ACADIA							•					•				•		•	•	•						
ARCHES																										
BADLANDS																										
BIG BEND																										
BISCAYNE								•								•		•		•	•	•				
BLACK CANYON OF THE GUNNISON																										
BRYCE CANYON																										
CANYONLANDS																										
CAPITOL REEF																										
CARLSBAD CAVERNS																										
CHANNEL ISLANDS		•	•	•	•			•			•		•			•	•	•	•	•	•		•	•	•	•
CONGAREE																										
CRATER LAKE																										
CUYAHOGA VALLEY																										
DEATH VALLEY																										
DENALI																										
DRY TORTUGAS																						•				
EVERGLADES								•														•				
GATES OF THE ARCTIC																										
GLACIER																										
GLACIER BAY							P	P			P		P			•	•	•	•	P	P			•		•
GRAND CANYON																										
GRAND TETON																										
GREAT BASIN																										
GREAT SAND DUNES																										
GREAT SMOKY MOUNTAINS																										
GUADALUPE MOUNTAINS																										
HALEAKALA																										
HAWAII VOLCANOES																		•								
HOT SPRINGS																										
ISLE ROYALE																										
JOSHUA TREE																										
KATMAI													•				•							•		•
KENAI FJORDS								P					P	•		•	•	•	•		•			P	P	•
KOBUK VALLEY														•												
LAKE CLARK																		•	•							•
LASSEN VOLCANIC																										
MAMMOTH CAVE																										
MESA VERDE																										
MOUNT RAINIER																										
NORTH CASCADES																										
OLYMPIC																	•	•	•					•	•	•
PETRIFIED FOREST																										
REDWOOD			•		•			•								•	•	•	•		•				•	•
ROCKY MOUNTAIN																										
SAGUARO																										
SEQUOIA–KINGS CANYON																										
SHENANDOAH																										
THEODORE ROOSEVELT																										
VIRGIN ISLANDS	P	P	P	P	•		•		P	•	P				P			•	P			P				
VOYAGEURS																										
WIND CAVE																										
WRANGELL–ST.ELIAS																	•	•	•						•	•
YELLOWSTONE																										
YOSEMITE																										
ZION																										

	Hair Seals							Foxes						Cats			Bears					Mustelids			
• species present P possible i introduced r reintroduced x extirpated	Harbor seal	Gray seal	Northern elephant seal	Sea otter	River otter	Coyote	Gray wolf	Arctic fox	Gray fox	Island fox	Red fox	Kit fox	Swift fox	Mountain lion	Bobcat	Lynx	Black bear	Grizzly bear	Raccoon	Ringtail	White-nosed coati	Least weasel	Long-tailed weasel	Short-tailed weasel	American marten
ACADIA	•	•			•	•					•				•		•		•				P	P	
ARCHES					•	•			•		•	•		•	•				•	•					
BADLANDS					•	•			•		•		•						•			•	•	•	
BIG BEND						•			•			•		•					•	•	P		•		
BISCAYNE					•				•										•						
BLACK CANYON OF THE GUNNISON					•	•	x		•					•					•	•			•		
BRYCE CANYON						•			•					•						•			•		
CANYONLANDS					•	•			•		•	•							•	•					
CAPITOL REEF				x	•	x			•		•	P		•	•		•		P	•				•	
CARLSBAD CAVERNS					•				•			•		•	•		•		•	•			•		
CHANNEL ISLANDS	•		•	•						•															
CONGAREE					•				•						•				•				•		
CRATER LAKE					•	•			•		•			•	•	•	•		•	P			•	•	•
CUYAHOGA VALLEY						•			•		•						•					•	•		
DEATH VALLEY						•			•			•		•	•				•						
DENALI					•	•	•				•					•	•	•					•		•
DRY TORTUGAS																									
EVERGLADES					•				•		•			•	•		•		•		i		•		
GATES OF THE ARCTIC					•	•	•	•			•				•	•	•	•				•			•
GLACIER					•	•	•				•			•	•	•	•	•				•	•	•	•
GLACIER BAY	•			•	•	•	•				•			•	•	•	•	•				•		•	•
GRAND CANYON					•	•			•		P	P		•	•		•		•	•			•		
GRAND TETON					•	•	r				•			•	•	•	•	•				•	•	•	•
GREAT BASIN						•			•			•		•	•				•				•	•	
GREAT SAND DUNES						•	x		•		•			•	•		•	x	•	P			•		P
GREAT SMOKY MOUNTAINS					r	•	x		•		•			x	•		•		•				•		
GUADALUPE MOUNTAINS						•	x		•				P	•	•		•	x	•	•			P		
HALEAKALA																									
HAWAII VOLCANOES																									
HOT SPRINGS					•	•			•		•				•		•		•				•		
ISLE ROYALE					•	x	•				•					x								•	P
JOSHUA TREE						•			•			•		•	•		•		•				•		
KATMAI	•			•	•	•					•				•		•					•		•	•
KENAI FJORDS	•		P	•	•	•	•				P				•	•	•							•	•
KOBUK VALLEY	•				•		•				•				•	•	•					P		•	•
LAKE CLARK	•			•	•	•	•				•			•	•	•	•					•		•	•
LASSEN VOLCANIC					•	•	x		•		•			•	•		•		•	•			•	•	P
MAMMOTH CAVE					r	•			•		•				•				•				•		
MESA VERDE						•	x		•		•	P		•	•		•		•	•			•		
MOUNT RAINIER						•					•			•	•		•		•				•	•	•
NORTH CASCADES					•	•	P				•			•	•	•	•	•				•	•	•
OLYMPIC	•		•	r	•	•	x				i			•	•		•		•				•	•	•
PETRIFIED FOREST						•	x		•		P	•		•	•		•	x	•	•					
REDWOOD	•		•		•	•			•		•			•	•		•		•				•	•	•
ROCKY MOUNTAIN					•	•	x		•		•			•	•	x	•	x	•				•	•	•
SAGUARO						•			•					•	•		•		•	•	•				
SEQUOIA–KINGS CANYON					P	•	x		•		•			•	•		•	x	•				•	•	
SHENANDOAH						•			•		•				•		•		•				•		
THEODORE ROOSEVELT					x	•					•		•	•	•		•		•			•	•		
VIRGIN ISLANDS																									
VOYAGEURS					•	•	•				•				P	P	•		•			P	P	•	•
WIND CAVE						•	x				P			•	•		x	x				P	•		
WRANGELL–ST. ELIAS	•			•	•	•	•				•					•	•	•				•		•	•
YELLOWSTONE					•	•	r				•			•	•		•	•	P				•	•	•
YOSEMITE					•	•			•		•			•	•		•	x	•	•			•	•	•
ZION						•	x		•		•	•		•	•		•	x	•	•			•	•	

208

Legend:

- **•** species present
- **P** possible
- **i** introduced
- **r** reintroduced
- **x** extirpated

	Mustelids (cont'd)					Skunks						Deer									Mt. Sheep		Exotic Mammals				
Park	Fisher	Black-footed ferret	Mink	Wolverine	Badger	Eastern spotted skunk	Hooded skunk	Striped skunk	Western hog-nosed skunk	Western spotted skunk	Collared peccary	Elk	Mule deer	White-tailed deer	Moose	Caribou	Pronghorn	Bison	Muskox	Mountain goat	Dall's sheep	Bighorn sheep	Wild pig	Burro	Wild horse	Small Indian mongoose
ACADIA	P		•					•						•	•											
ARCHES			•					•		•			•				•					•				
BADLANDS		r			•	•		•					•	•			•	r				r				
BIG BEND			•				P	•	•	•	•	•	•	•			•					r	P			
BISCAYNE								•						•												
BLACK CANYON OF THE GUNNISON			P		•			•				•	•									•				
BRYCE CANYON			•					•		•		•	•				•									
CANYONLANDS			•					•		•			•				•					•				
CAPITOL REEF		•			•			•		•			•				x	r				x				
CARLSBAD CAVERNS			•					•	•	•	r	i	•				•	x				x				
CHANNEL ISLANDS										•																
CONGAREE			•					•						•									•			
CRATER LAKE	•		•	P	•			•				•	•	P			•									
CUYAHOGA VALLEY			•					•						•												
DEATH VALLEY					•					•			•									•		•	•	
DENALI			•	•											•	•				P	•					
DRY TORTUGAS																										
EVERGLADES			•			P		•						•									•			
GATES OF THE ARCTIC			•	•											•	•			•		•					
GLACIER	•		•	•	•			•				•	•	•	•					•		•				
GLACIER BAY			•	•									•		•					•						
GRAND CANYON			•					•		•		•	•				•					•		P		
GRAND TETON		•	•		•			•				•	•	•	•		•	•		P		•				
GREAT BASIN			•					•		•		r	•				•					r			P	
GREAT SAND DUNES		x		P	•			•		P		•	•				•	x				•				
GREAT SMOKY MOUNTAINS	x		•			•		•				r		•				x					•			
GUADALUPE MOUNTAINS					•			•	•	•	•	i	•				x	x				x				
HALEAKALA																							•		•	•
HAWAII VOLCANOES																							•			•
HOT SPRINGS			•			•								•												
ISLE ROYALE			•					•						x	•	x										
JOSHUA TREE					•			•		•			•									•				
KATMAI			•	•											•	•										
KENAI FJORDS			•	•											•					•	•					
KOBUK VALLEY			•	•											•	•					•					
LAKE CLARK			•	•											•	•					•					
LASSEN VOLCANIC	P		•		•			•		•		•	•				•									
MAMMOTH CAVE			•			•		•						r												
MESA VERDE		x			•			•		•		•	•	x								r			•	
MOUNT RAINIER	•		•					•		•		•	•							•						
NORTH CASCADES	•		•	•	•			•		P		•	•	P	•					•		P				
OLYMPIC	P		•					•		•		•	•							i				•		
PETRIFIED FOREST					•												•									
REDWOOD	P		•									•	•										•			
ROCKY MOUNTAIN			•	P	•			•		•		•	•	•	•					x		r				
SAGUARO					•		•	•	•	•	•		•													
SEQUOIA–KINGS CANYON	•		P	P	•			•		•			•									•	•			
SHENANDOAH			•			•		•						•									•			
THEODORE ROOSEVELT		x	•		•			•				r	•	•			•	r				r			•	
VIRGIN ISLANDS																							•	•		•
VOYAGEURS	•		•		P			•				x		•	•	x										
WIND CAVE		x	P		•	P		•				r	•	•			r	r								
WRANGELL–ST.ELIAS			•	•											•	•		i		•	•					
YELLOWSTONE	P		•	•	•			•				•	•	•	•		•	•		i		•				
YOSEMITE	•		•	P	•			•		•		•	•									•	•			
ZION					•			•		•		•	•									r				

209

PHOTOGRAPHY CREDITS

Photographs by the following photographers appear on the pages listed:
Jim Brandenburg (Minden Pictures), xii–1, 20–21, 34, 46, 53, 55, 63, 84–85, 111, 165, 181, 183, 190; John Burde, 37, 49, 79; Gerry Ellis (Minden Pictures), 117; Tim Fitzharris (Minden Pictures), 17, 40–41, 52, 70–71, 93, 95, 97, 150, 162, 176; Michael & Patricia Fogden (Minden Pictures), 115; Sumio Harada (Minden Pictures), i, 56–57, 107, 123, 126, 132, 174, 175, 194; Michino Hoshino (Minden Pictures), 36, 60, 144, 188–189, 193; Heidi & Hans-Juergen Koch (Minden Pictures), 108–109; Frans Lanting (Minden Pictures), 153; Sarah Leen, vii (top), 6–7; Thomas Mangelsen (Minden Pictures) xi, 172; Bill Marr, 80; Yva Momatiuk & John Eastcott (Minden Pictures), 156–157, 199; Flip Nicklin (Minden Pictures), 131, 140, 143, 146–147; Michael Quinton (Minden Pictures) viii, 30–31, 102, 112, 136, 139, 171, 179; Joel Sartore, ii–iii, vii (bottom), 2–3, 4–5, 98–99, 128, 149, 152, 158, 166, 186, 191, 196; Merlin Tuttle (Bat Conservation International), 104 (top and bottom); Tom Vezo (Minden Pictures), 61, 122; Konrad Wothe (Minden Pictures), 100, 120–121, 134–135, 154, 169, 177, 178; Norbert Wu (Minden Pictures), 106; Shin Yoshino (Minden Pictures), 125. Photographs courtesy of the following sources appear on the pages listed: National Park Service, 8, 10, 11, 12, 14, 15, 16, 18, 19, 22, 25, 27, 33, 39, 44, 47, 51, 58, 64, 67, 69, 72, 73, 75, 76, 77, 81, 83, 87, 88, 89, 91; U.S. Fish & Wildlife Service, 137, 145, 163, 173, 182, 185.